Neural Pathways to Emotional Mastery

By
Eliora B. Mindwell

Neural Pathways to Emotional Mastery

Table of Contents

Introduction

In the bustling modern world, professional and personal landscapes are evolving more swiftly than ever before. At the heart of this transformation is a quality that, while ancient in its presence, is only now being deeply understood in scientific terms: emotional intelligence. This book serves as a bridge between the hard science of neuroscience and the soft skills of emotional intelligence, offering you the tools to forge stronger connections, make wiser decisions, and enhance both personal and professional life through a clearer understanding of your own brain.

Emotional intelligence, in its simplest form, is the ability to be aware of, control, and express one's emotions while also handling interpersonal relationships judiciously and empathetically. It's a skill set as crucial as it is nuanced. Despite its complexity, emotional intelligence can be broken down and understood with the help of neuroscience, offering everyone the chance to improve and hone these essential skills. This isn't just about better knowing yourself; it's also about navigating the world around you with greater confidence and compassion.

Why is now the right time to focus on this journey of understanding? As technology propels us into new realms of connectivity, the paradox of modern life emerges: we are connected yet often feel isolated; we communicate more yet understand less. Neuroscience offers a beacon of hope amidst this paradox, guiding us toward a deeper comprehension of our emotional selves and thus charting paths to more fulfilling interactions. Utilizing this knowledge encourages us to tap into parts of our potential we might've never realized were there.

This introduction opens a door to a comprehensive exploration of how our brain shapes our emotions, decisions, interactions, and ultimately, our lives. We'll uncover how neural pathways mold our responses, influence our stress levels, and even alter our capacity for empathy. You'll gain insights into the subtle dance between neurochemicals and their impact on mood and motivation, equipping you with the knowledge to influence your emotional state proactively.

The chapters that follow will not just scrub the surface but delve deeply into how you can rewire your brain to enhance emotional intelligence, an asset that pays dividends in every facet of life. You might wonder: can people truly change the way they feel and respond? The answer, rooted in the concept of neuroplasticity, resounds with a confident yes. No longer just a topic for scientific circles, neuroplasticity is a powerful tool you can harness for emotional growth and resilience.

A key theme throughout this book will be the practical application of scientific insights. Theory without practice is hollow, and so this book intends to equip you with actionable strategies for cultivating resilience, managing stress, and fostering empathy. Whether you're in the boardroom or the living room, these skills are invaluable. They form the fabric of meaningful relationships and effective leadership, intertwining personal satisfaction with professional success.

Moreover, you'll discover the importance of self-care in emotional intelligence. Recognizing when and how to seek moments of mindfulness, the value of sleep, nutrition, and physical activity, all act as pillars supporting emotional well-being. Science has shown us that these aren't mere habits but vital components to maintaining the balance required for emotional health and cognitive longevity.

Considering the rapid pacing of technological advancement, another fascinating aspect this book will address is the intersection of technology and emotion. While our devices can pull us away from the tangible world, they also hold the potential to enhance our emotional intelligence, provided we use them wisely. We'll explore innovative ways to leverage technology for emotional enhancement without becoming enslaved by it.

The groundbreaking realm of neurotherapy also makes its presence felt in our exploration. New therapeutic approaches offer exciting possibilities for emotional healing, showcasing a future where the brain's profound capacity to heal and adapt is fully embraced and utilized. It represents a growing frontier in both personal transformation and mental health.

As we voyage through these ideas, remember that developing emotional intelligence isn't a destination but a journey, one you'll refine and redefine throughout your lifetime. It's an invitation to perpetual growth, guided by the expanding insights from neuroscience. However, the potential is already within you, waiting to be unlocked.

In closing, this book is your roadmap for personal evolution, a guide to turning insights into action, and words into wisdom. As you engage with each chapter, let your curiosity lead, and let your newfound understanding of the brain inspire transformative changes in how you live, love, and lead. Let's embark on this journey of discovery and empowerment, as you harness the power of neuroscience to enrich both your inner life and the world you inhabit.

Chapter 1:
The Neuroscience of Emotions

Emotions shape our personal and professional lives in profound ways, influencing decisions, relationships, and mental health. At the heart of understanding emotions lies the intricate network of our brain, where complex processes determine how we experience, express, and manage emotional responses. This chapter delves into the neuroscience that underpins these processes, offering insight into how key structures in the brain, such as the amygdala and prefrontal cortex, play significant roles in emotional processing. While emotions might often be dismissed as intangible or abstract, neuroscience reveals that they are anything but; rather, they are rooted in physiological processes that can be studied, understood, and influenced. By exploring the science behind emotions, professionals can better harness their emotional landscape, leading to enhanced interpersonal skills, decision-making, and even leadership capabilities. This understanding paves the way for harnessing emotions to foster both personal and collective growth, transforming challenges into opportunities through increased emotional intelligence.

Understanding Emotional Processing in the Brain

Emotions are intricately woven into our humanity, a complex tapestry that colors our daily lives. Yet, understanding how our brains process these feelings remains a sophisticated puzzle that scientific minds have been striving to piece together. Our journey into this realm is not just a mere exploration of physiology but a quest to harness this knowledge for profound personal and professional growth. Emotions influence our decisions, shape our relationships, and define our sense of self. Recognizing the mechanisms that underlie emotional processing empowers us to cultivate emotional intelligence and deploy it as a transformative tool.

In the neural orchestra playing the symphony of emotions, certain brain areas act as the primary conductors. Though several regions participate in this intricate operation, the limbic system is often the focal point of discussions. It plays a central role in regulating emotions and consolidating memories. Within the limbic system, the amygdala, often dubbed the 'emotion center,' stands out. This almond-shaped cluster of nuclei is crucial for the perception of emotions such as fear, pleasure, and anger. It helps tag emotional significance to memories, preparing the individual to respond appropriately to similar stimuli in the future.

Adjacent to the amygdala lies the hippocampus, a structure often associated with the formation of new memories. However, its influence on emotions shouldn't be understated. It works alongside the amygdala, providing contextual background—the who, what, and when that accompanies emotional events. Imagine watching a dramatic scene unfold in your favorite movie; the hippocampus is busy cataloging the storyline, while the amygdala attaches emotional responses to each climactic moment. Together, these structures not only affect how we relive memories but also shape how we react to new experiences.

The prefrontal cortex (PFC) acts as a sophisticated moderator in this emotional framework. Positioned at the brain's forefront, it is associated with complex cognitive behavior and decision-making. Its involvement in emotional regulation highlights the nexus between cognition and emotion. The PFC evaluates emotional inputs, modulates responses, and orchestrates a balanced reaction. Have you ever counted to ten when angry or paused to think before speaking? That's the PFC in action, tempering your amygdala's impulsive emotional output with rational oversight.

Yet, emotional processing isn't solely fixed within these structures. It emerges from dynamic interactions across various brain regions, each contributing different dimensions to the emotional experience. The brain uses vast networks and pathways, like highways connecting bustling cities, to integrate sensory and cognitive information. This interconnectedness ensures emotions are neither stagnant nor isolated; they're fluid, contextual, and adaptable.

The insular cortex, often overshadowed by more prominent structures, merits attention for its role in emotional awareness and empathy. It orchestrates the subjective emotional experience by integrating visceral and emotional signals, acting as a bridge between the emotional and bodily sensations. Consider the visceral responses of excitement: the racing heart, the tingly skin. The insular cortex processes these signals, contributing to your awareness of your emotional state.

Neuroscience reveals that even the perception of sounds, sights, and smells can activate emotional circuits. The anterior cingulate cortex (ACC), for example, plays a crucial role in error detection and anticipation of outcomes, linking emotional processes with attention and cognitive control. It is a monitoring hub, scanning for deviations from the norm and determining the emotional salience of these changes. When faced with a novel situation, the ACC assesses potential threats or rewards, modulating emotional responses accordingly.

As we continue to untangle the intricacies of emotional processing, it's evident that neurochemicals and hormones serve as the brain's chemical messengers, further influencing our emotional landscape. Serotonin, dopamine, and cortisol are central players in this arena, modulating mood, pleasure, and stress responses respectively. These chemicals circulate through specific pathways and synapses, creating a symphony of signals that underpin emotional experiences.

Understanding the brain's emotional processing is not just a cerebral exercise—it's a pathway to emotional mastery. By recognizing patterns and responses etched in our brain's circuitry, we become equipped with the means to hack our emotional systems purposefully. It's like acquiring a map of a complex terrain; with this map, we can navigate our emotional world more effectively, fostering deeper empathy, resilience, and well-being.

Consider the practical applications of this knowledge. In stressful situations, awareness of the brain's emotional processing mechanisms allows for strategic interventions. Techniques like mindfulness and cognitive restructuring can be employed to recalibrate the prefrontal-limbic balance, fostering emotional regulation and resilience. Such strategies capitalize on the brain's inherent neuroplasticity, sometimes likened to a sculptor molding clay, reinforcing neural pathways that support adaptive emotional responses.

Moreover, understanding emotional processing aligns with broader goals of enhancing emotional intelligence within professional realms. Imagine a leader who grasps these nuances—a leader who reads emotional cues with precision, responds with empathy, and builds rapport effortlessly. Such leaders leverage their understanding of the brain's emotional dynamics to create environments where creativity flourishes and collaboration thrives.

The exploration of emotional processing in the brain is an invitation to delve into the human condition's very essence. It's a journey through the labyrinth of thought and feeling, a blend of rational analysis and emotional intuition. While science continues to unlock new insights, the lessons drawn from current understanding provide a powerful toolset for personal empowerment and societal transformation.

In closing, the cerebral dance of emotional processing challenges us to redefine our relationship with our emotions. Armed with knowledge, we can engage more fully with ourselves and others, paving the way for growth that's deeply rooted in the wisdom of our neurology. It is a call to action— an opportunity to become not only observers of our responses but also active participants in the ever-evolving narrative of our lives.

Key Brain Structures Involved in Emotion

The intricate tapestry of human emotion has long fascinated scientists, philosophers, and self-discovery enthusiasts alike. At the heart of this enigma lies the brain—a marvel of biological engineering. This section delves into the neural architecture that underpins our emotional lives. Understanding these key brain structures equips us with the knowledge to cultivate emotional intelligence, ensuring we're not just reactive creatures of instinct but proactive stewards of our emotional well-being.

The *amygdala* often takes center stage in discussions about emotions. Known as the brain's emotional watchdog, it scans incoming information for threats and rewards, triggering emotional responses that are often visceral and immediate. Despite its small size, the amygdala plays a monumental role in how we process emotions like fear and pleasure. When you suddenly feel an uptick of anxiety or excitement, it's likely the handiwork of this almond-shaped cluster of nuclei. Its operations don't just affect how we feel; they influence how we perceive and react to the world around us. Understanding its mechanisms can empower us to modulate our responses and reduce emotional reactivity.

Beneath the cerebral cortex lies another crucial component called the *hippocampus*. Known mainly for its role in memory formation, the hippocampus intertwines closely with the amygdala to assign emotional significance to memories. Imagine recalling a bittersweet memory from your past—the warm nostalgia intermingled with a twinge of sadness. While the amygdala colors your emotional response, the hippocampus anchors this reaction in memory, ensuring that significant events leave a lasting imprint. This association underscores how memory and emotion are fundamentally interrelated, highlighting the importance of nurturing positive experiences.

Directing traffic in the bustling cityscape of our brain is the *prefrontal cortex*—often hailed as the seat of executive functions. Here, impulses are weighed, decisions contemplated, and emotions regulated. In the dance of emotional response, the prefrontal cortex partners with both the amygdala and hippocampus, providing a rational counterweight to impulsive emotional reactions. It's the locus of our capacity for self-control, enabling us to pause and reflect rather than act out of raw emotion. This brain region is vital in practices like mindfulness management and emotional intelligence enhancement, as it transforms immediate reactions into thoughtful responses.

Meanwhile, the *insula* plays its own distinct yet interconnected role in how we experience emotions. Bridging our physical sensations with emotional experiences, the insula is responsible for the feeling of 'gut instincts.' When confronted with moral dilemmas or empathizing with another person, it creates a visceral reaction that informs our moral compass. Understanding this introspective brain region offers the key to tuning into our internal states and improving our emotional self-awareness.

Another elemental player in emotion is the *anterior cingulate cortex (ACC)*. Situated close to the prefrontal cortex, the ACC integrates emotional and cognitive processing. Acting as an emotional compass, it directs attention and deploys mental resources in response to emotional stimuli. When you feel conflicted about a decision, it's likely your ACC working overtime to weigh emotions against rational thought. Enhancing its functionality can thus support emotion-driven decision-making, enabling actions that align with our deeper values and goals.

Engaging in a silent dialogue with these components is the *hypothalamus*, which links the nervous system to the endocrine system. Among its many roles, it regulates stress responses, releasing hormones like cortisol when you feel threatened. However, in today's fast-paced world, where the line between a real threat and everyday stressors gets blurred, the hypothalamus can sometimes become overactive. Strategies that help balance its output—such as stress management and lifestyle adjustments—can prevent the burnout often resultant from chronic stress.

The *ventromedial prefrontal cortex (vmPFC)*, meanwhile, modulates social emotions such as guilt, embarrassment, and compassion, impacting our ability to relate to others. This region helps integrate emotional signals with the knowledge of social norms, guiding our interactions. A robust vmPFC function underlies empathetic communication and emotional regulation, both crucial skills in personal and professional relationships.

Games of emotions often involve a large cast, notably including *mirror neurons*. These neurons, found primarily in the premotor cortex, fire both when we perform an action and when we see someone else perform the same action. They are critical for imitation learning and empathy—essentially allowing us to 'feel' what others are going through. This biological empathy tool exemplifies how our brain is hardwired for social connection, underscoring the potential to harness it for enhanced interpersonal experiences.

Understanding these key brain structures and their interactions isn't solely an academic pursuit. When we learn how to balance the conversation between our amygdala and prefrontal cortex, we're not just managing stress—we're enhancing our emotional intelligence. By fostering neural pathways that support resilience, we build internal fortresses capable of weathering life's storms. Empowered with this knowledge, the journey into emotional mastery isn't just possible—it's within our reach, a testament to the remarkable synergy of malleable minds engaged in the lifelong dance of learning and evolution.

In our quest for emotional empowerment, the neuroscience of emotions offers a roadmap, guiding us through the complexities of our emotional landscapes. By leveraging these insights, we cultivate a deeper understanding of ourselves and others, creating pathways towards a more harmonious existence. The challenge then is not merely to possess this knowledge but to apply it, transforming our lives—and the lives of those around us—into richer, more connected experiences.

Chapter 2:
Neuroplasticity and
Emotional Growth

As we transition from understanding the neuroscience of emotions, we enter the dynamic world of neuroplasticity—a key to unlocking emotional growth. This fascinating concept reveals our brain's ability to adapt and reorganize itself throughout our lives, offering profound opportunities for personal transformation. Neuroplasticity doesn't just stitch a new pattern into our neurons; it opens doors to alter emotional experiences and responses, enhancing our emotional intelligence incrementally. By intentionally shaping our neural pathways, we lay the groundwork for more profound empathy, resilience, and emotional regulation. Through deliberate practice and focused techniques, such as mindfulness and cognitive restructuring, individuals can effectively rewire their emotional responses. This isn't just brain science; it's a call to harness your brain's potential for a richer, more emotionally intelligent existence. As we delve further into this journey, remember that this transformative power lies within your grasp, guiding you toward a future where emotional growth becomes a lived reality.

Harnessing Neuroplasticity for Emotional

Development

In the landscape of personal growth, the idea that our brains are not static but rather malleable canvases comes as a revelation. This concept, known as neuroplasticity, illuminates a pathway to developing our emotional capacities with intention and insight. Neuroplasticity underscores the brain's ability to reorganize itself by forming new neural connections throughout life, which can significantly impact how we perceive and manage our emotions.

The profound potential of neuroplasticity becomes even more impactful when we realize that it's not confined to a specific stage in life. From childhood through adulthood, our brains retain a remarkable ability to adapt to new experiences, change patterns, and heal from emotional adversities. This adaptability offers a foundation for emotional development. You don't just change the way you react; you transform the ever-evolving neural circuits that give rise to your emotional responses.

Consider the way we learn to play a musical instrument. The first notes are often clumsy, awkward, and disjointed. With practice, however, something magical happens. The neural circuits in our brain reorganize, strengthening the connections related to musical coordination and auditory processing. Emotional development works much the same way. With conscious effort and practice, we can reshape our brain's response to emotional stimuli, fostering resilience, empathy, and emotional intelligence in the process.

The science behind this transformation is both fascinating and encouraging. Neuroplasticity operates through mechanisms like long-term potentiation, where synaptic connections are strengthened, and neurogenesis, the birth of new neurons. These processes can be activated and optimized through experiences, thoughts, and behaviors that stretch your emotional faculties. Rather than being fixed, your emotional intelligence can grow and expand, paralleling your increasing understanding of neural dynamics.

Practical application is at the heart of harnessing neuroplasticity. Techniques such as mindfulness, deliberate practice, cognitive-behavioral interventions, and even artistic engagement stimulate brain plasticity. When you mindfully engage with your emotions, observing them without judgment, you cultivate a non-reactive awareness that can alter your brain's default pathways. This practice not only calms the amygdala, which is involved in fear and stress responses, but also strengthens the prefrontal cortex, enhancing decision-making and emotional regulation.

Deliberate practice in emotional scenarios—such as empathizing with others, reinterpreting negative emotions, or even envisioning positive resolutions to conflicts—can leave lasting imprints on your neural structure. By repeatedly engaging in these exercises, you're not just viewing transformations in your emotional responses; you're witnessing a physical change in your brain circuitry.

Imagine a situation where previously, anger might have flared up in response to criticism. Through targeted emotional exercises and awareness, you can gradually rewire your brain to interpret criticism as an opportunity for growth rather than a personal attack. This shift doesn't occur overnight, but with continued effort, it becomes a tangible reality—thanks to neuroplasticity.

Incorporating emotional development into your routine could also involve engaging in artistic endeavors like painting or music. Artistic expressions not only stimulate creativity but also fortify emotional growth. The act of creating art serves as both a mirror to reflect existing emotional states and a canvas to project and transform emotions, all of which contributes to restructuring neural pathways.

The ultimate aim of harnessing neuroplasticity is not to control our emotions artificially but to establish healthier patterns and reactions that contribute to overall well-being. By fostering emotional development through understanding and applying the principles of neuroplasticity, we gain the tools to navigate the complexities of human emotion with greater ease and flexibility.

This journey of emotional enhancement requires patience, self-compassion, and persistence. You might falter at times or confront resistance from well-worn neural pathways. But with each effort, you edge closer to the potential for a balanced and enriched emotional life. The adaptability of your brain is a powerful ally in this quest.

In a world that often equates intelligence with cognitive abilities alone, neuroplasticity reminds us that emotional intelligence is equally attainable and just as crucial. It invites us to participate actively in our emotional evolution, empowering us to pursue not only personal growth but also richer, more meaningful connections with the world around us.

Embracing the promise of neuroplasticity is akin to embarking on a journey of self-discovery and transformation. As you practice, adapt, and grow, remember that each step you take is a testament to the dynamic nature of your brain and its capacity for infinite growth. Through the lens of neuroscience, you're not just learning to manage your emotions—you're learning to master them.

Practical Techniques for Emotional Rewiring

As we delve into the realm of neuroplasticity, the ability of the brain to reorganize itself by forming new neural connections, we unearth powerful tools that can dramatically reshape our emotional landscape. The concept of neuroplasticity is not just a scientific curiosity; it is a beacon of hope for those seeking to enhance their emotional intelligence and interpersonal skills. This adaptability allows us to rewire our emotional responses, turning challenges into stepping stones for personal growth. So, what practical techniques can we adopt to harness this incredible capability?

One effective method for emotional rewiring is **mindfulness meditation**. By engaging in regular mindfulness practices, individuals can increase their awareness of emotional triggers and responses, creating a space between stimulus and reaction. This pause is essential for developing thoughtful and intentional emotional responses rather than knee-jerk reactions. Through consistent practice, the brain begins to develop new pathways that promote calmness and emotional resilience, empowering us to handle stress more effectively.

Gratitude practice is another technique that plays a significant role in emotional reshaping. Making a habit of identifying and appreciating the positive aspects of our lives can dramatically alter brain chemistry. When practiced regularly, gratitude can enhance mood by fostering a positive emotional outlook and increasing levels of the neurotransmitters serotonin and dopamine, which are linked to feelings of happiness and satisfaction. This positive feedback loop encourages further engagement with life, highlighting the interconnectedness of mindset and brain structure.

For those looking to tackle more challenging emotions, **emotional reframing** offers a robust tool. This technique involves consciously changing the narrative or context of an emotional experience. By altering our perception of a situation, we can change our emotional response. For instance, perceiving a difficult interaction as an opportunity for growth rather than a threat can transform fear into excitement or confidence. Reframing helps lay down new neural patterns that support positive emotional shifts.

Another practical approach is the use of **cognitive behavioral techniques (CBT)**. These methods focus on identifying and altering unhealthy thought patterns that lead to emotional difficulties. By addressing distorted thinking, individuals can reduce stress, depression, and anxiety, thereby modifying their emotional responses. The brain's plasticity ensures that as we introduce healthier cognitive patterns, they gradually replace the less adaptive ones, leading to more stable emotional health over time.

Additionally, **emotional awareness exercises** can be invaluable. These exercises, which include journaling, consciously tracking emotions throughout the day, and reflective exercises, help individuals identify patterns and triggers in their emotional responses. By understanding these elements more deeply, it becomes easier to redirect negative emotions and reinforce positive ones, harnessing neuroplasticity to enhance emotional well-being.

For those who thrive on interpersonal interaction, **social support systems** provide an indispensable resource for emotional development. Engaging with social groups, whether through family, friends, or professional networks, helps build emotional resilience. Secure social connections foster a nurturing environment where neural pathways related to empathy, compassion, and emotional intelligence are strengthened. The positive emotions generated through these interactions bolster mental health and fortify the brain against stressors.

Finally, incorporating **physical activity** into daily routines cannot be understated. Exercise has a profound effect on brain plasticity, promoting the release of neurochemicals such as endorphins, which are natural mood lifters. Regular physical activity not only reshapes the body but also invigorates the mind, enhancing its capacity to handle emotions. When combined with the above techniques, exercise provides a multifaceted approach to emotional rewiring.

The journey to rewiring our emotions through neuroplasticity is not merely about eliminating negative emotions but rather about cultivating an emotional environment where both challenging and joyous emotions can be understood, managed, and even celebrated. By employing these practical techniques, we not only enhance our emotional intelligence but also foster a more profound connection with ourselves and those around us. The empowerment that results from taking control of our emotional growth is not just transformative but deeply liberating.

Chapter 3:
The Science of Emotional Intelligence

Emotional intelligence (EI) is a cornerstone of personal and professional success, blending cognitive processes with nuanced emotional understanding. It's not just about being emotionally aware but also about harnessing this awareness to navigate social complexities and foster thriving relationships. The brain, a marvel of interconnected systems, plays a vital role in EI by bridging emotion with intellect. Understanding this connection allows us to break down emotional intelligence into tangible components—like self-awareness, empathy, and emotional regulation—that can be enhanced through practice and neuroscience. By delving into the intricate dance between the emotional and cognitive faculties, professionals and lifelong learners can unlock new dimensions of personal growth. This pursuit isn't merely academic; it's a powerful, transformative journey into the heart of what makes us human, driving us to not only perceive the emotions of others but to respond with kindness, understanding, and strategic thinking. Bridging the realms of emotion and logic, the science of emotional intelligence empowers individuals to build resilience, adaptability, and interpersonal acuity, becoming not just more competent professionals but more insightful and compassionate individuals. With each layer of this complex interplay that we uncover, we gain the tools to lead with empathy and innovate with emotional acuity, paving the path for a more connected and emotionally intelligent world.

Breaking Down Emotional Intelligence

Emotional intelligence (EI) serves as the cornerstone for many of the skills discussed in this book, providing the framework for understanding how we perceive, control, and express our emotions. It's not just an abstract concept; it's a tangible skill set that can be honed over time. The theory of emotional intelligence encompasses various abilities, including recognizing, understanding, and managing our own emotions, while also being able to recognize, understand, and influence the emotions of others. These components work together to equip individuals with the tools they need to handle social complexities and make informed, empathetic decisions.

Breaking down emotional intelligence starts with self-awareness. This ability forms the very bedrock of EI, allowing individuals to comprehend the nuances of their own emotional landscapes. Self-awareness involves an honest evaluation of one's feelings and their impact on decision-making and relationships. It's about acknowledging emotions without being overwhelmed by them. In essence, self-awareness acts as a mirror that reflects our innermost thoughts and feelings, enabling us to address them calmly and constructively.

Following self-awareness, self-regulation comes into play. This involves the discipline to control or redirect disruptive emotions and impulses and to think before acting. With self-regulation, you maintain flexibility and adaptability to change. It's about channeling your emotional energy into productive outlets, ensuring that your reactions are appropriate to the situations you face. Imagine navigating a turbulent sea; self-regulation is the rudder that helps steer your response in the right direction, keeping you on course even amidst emotional upheaval.

Another critical element of EI is motivation, which serves as the internal drive that propels individuals toward achieving their goals. Beyond monetary reward or social recognition, motivation in the context of emotional intelligence is intrinsic. It's about striving for self-improvement, setting high standards, and persisting in the face of obstacles. Those high in emotional intelligence use personal goals to fuel their ambitions, drawing on their emotions as a reservoir of resilience and inspiration.

Social awareness, like a finely tuned sensor, picks up on the emotions of others. It's the ability to empathize, to put yourself in another's shoes, and to understand the emotional dynamics at play in various social settings. Social awareness enables individuals to recognize the needs of others, paving the way for more harmonious and effective interactions. This aspect of EI involves more than just intellectual understanding; it's about emotional attunement and genuine empathy, allowing for deeper connections with others.

The final piece of the emotional intelligence puzzle is relationship management. This skill revolves around the ability to nurture and maintain healthy relationships, drawing on all the other aspects of emotional intelligence. Whether leading a team through a challenging project or resolving a conflict, strong relationship management skills require negotiation, communication, and collaborative abilities. It's the art of influencing others while maintaining empathy, trust, and respect.

In the professional realm, emotional intelligence is often the differentiator between good and great leaders. Leaders with high EI can inspire and engage employees, fostering environments of trust and creativity where individuals are empowered to perform at their best. They don't merely react to situations; they respond with a strategic flexibility that considers the emotional tone of their team. In workplaces where emotional intelligence is recognized and valued, employees tend to experience higher job satisfaction, lower stress levels, and a greater sense of belonging.

Moreover, emotional intelligence plays a vital role in personal growth. By understanding your own emotions and those of others, you become better equipped to manage your relationships with family and friends. The insight gained from practicing EI encourages open communication, reduces conflicts, and builds stronger connections. You learn to navigate the complexities of human interactions with grace and understanding, turning potential conflicts into opportunities for growth and resolution.

Developing emotional intelligence is an ongoing journey. It's not about achieving perfection but about striving to improve continually. The essence of EI lies in its practice, in the small, everyday moments where we choose understanding over judgment, patience over haste, and empathy over indifference. Incorporating emotional intelligence into your life doesn't mean ignoring rational thought; instead, it enhances logical processes by integrating them with emotional awareness, delivering a comprehensive approach to challenges big and small.

In conclusion, breaking down emotional intelligence reveals a multi-faceted construct that significantly impacts both personal and professional realms. By enhancing emotional intelligence, you cultivate a toolbox of skills that equip you to handle life's challenges with poise and confidence. This ongoing journey empowers you to interact with the world in a more meaningful and productive way, guiding you toward not just accomplishing your goals, but doing so with empathy and insight. So, embark on this transformative path and watch as your understanding of emotions enhances every aspect of your life.

The Brain-Emotion-IQ Connection

Emotional intelligence, a term now embedded in the fabric of modern professional and personal spheres, rests on the intricate interface between our brain, emotions, and intellect. This connection is fundamental for anyone seeking to navigate the complexities of human interaction effectively. Our understanding of this triad is not just a theoretical pursuit but a practical guide to reshaping our interactions and enhancing our decisions across every facet of life.

The roots of emotional intelligence lie in the very machinery of our brain. Our emotional experiences are not mere responses; they are integral components of our cognitive processes. The brain, a marvel of biological engineering, controls emotions that influence feelings and drive actions. Central to this is the limbic system, a collection of structures such as the amygdala and hippocampus, which processes emotions and stores emotional memories. Meanwhile, the prefrontal cortex plays the role of manager, balancing these emotions with rational thought.

Understanding this balance is key to mastering emotional intelligence. The amygdala is known for triggering immediate emotional reactions, particularly those tied to survival like fear and pleasure. However, without the tempering influence of the prefrontal cortex, these reactions might lead to rash decisions or impulsive behavior. This dynamic interaction forms the bedrock of our emotional intelligence by allowing us to assess situations more comprehensively and managing our responses accordingly.

Our intellectual capabilities and our emotional responses frequently engage in a delicate dance. Take decision-making, for example. While we might like to think our choices are based solely on logic, emotions undeniably color our perceptions and judgments. Studies show that individuals who can recognize and understand their emotional biases tend to make better decisions, underscoring the critical role of emotions in cognitive processes.

Moreover, emotional intelligence is not static. Thanks to the concept of neuroplasticity, we know our brains can change and adapt throughout our lives. Engaging in practices that enhance emotional awareness, such as mindfulness, meditation, and reflective thinking, can significantly improve our emotional intelligence. These practices encourage the brain to create new pathways and connections, leading to improved emotional processing and regulation. Essentially, we can rewire our brain to foster better emotional responses and enhance our cognitive functions simultaneously.

The application of emotional intelligence in the professional realm is transformative. Leaders with high emotional intelligence tend to foster environments of trust and collaboration. This is achieved through heightened empathy, understanding, and communication skills, which create more harmonious and productive workplaces. Such leaders are adept at reading the emotional currents within their teams and responding in ways that align with their organizational goals. Whether dealing with the stress of a looming deadline or navigating interpersonal conflicts, the emotionally intelligent leader can steer towards solutions with composure and insight.

But it is not just in professional settings where the brain-emotion-IQ connection thrives. Personal relationships benefit hugely from this understanding. Empathy, arguably the cornerstone of emotional intelligence, allows individuals to connect on a more profound level, facilitating healthier, more meaningful interactions. This connection arises from the brain's ability to simulate others' experiences internally. When harnessed correctly, this can lead to greater understanding and stronger bonds.

Educational spheres are also ripe for revolution through enhanced understanding of this triad. By integrating insights into how emotions affect learning, educators can tailor their approaches to maximize student engagement and retention. It's a paradigm shift that recognizes education as not just imparting knowledge, but also fostering emotional and cognitive growth.

It is an exciting time as we uncover more about the neurological underpinnings of emotional intelligence. Yet, as scientific understanding deepens, we are reminded of humanity's timeless intuition: that emotions and intellect are not mutually exclusive but are deeply intertwined. The challenge is to leverage this knowledge, not for personal gain alone, but to contribute positively to our communities and the world at large.

This journey into understanding the brain-emotion-IQ connection is not merely academic. It is about empowering ourselves with the knowledge and tools necessary for personal growth and professional excellence. By understanding how our brains process emotions and how we can improve upon this, we embark on a path to greater emotional resilience, profound self-awareness, and ultimately, transformative personal and professional experiences.

With each step forward, we equip ourselves better to handle the complexities of life. Whether through investing time in reflective practices to bolster our emotional intelligence or adopting strategies to manage stress more effectively, the effort pays dividends. It leads to a more profound understanding of oneself and others and a life lived with greater intention and connection.

Chapter 4:
Understanding and
Managing Stress

In the perpetual hustle of professional life, understanding stress isn't just beneficial—it's essential. Our brain is hardwired to react to stress, but recognizing how these responses manifest can empower us to regain control. Stress isn't just a mental game; it impacts our physical well-being, driving what neuroscientists call the "fight or flight" response. This chapter delves into the nuances of stress, unveiling the brain's complex chemistry and circuitry designed to handle threats, whether real or perceived, with precision. Grasping these intricate mechanisms allows us to employ brain-based strategies to manage stress effectively. Techniques that foster neural reconfiguration can optimize our resilience, transforming stress from an adversary into an ally. By weaving scientific insights with practical interventions, we can cultivate a state of poised calm, enhancing both personal and professional arenas. This journey not only promises reduced anxiety but also nurtures a more fulfilling and balanced existence, where stress management is seamlessly integrated into our daily narrative.

Neuroscience Behind Stress Responses

Stress is an inevitable aspect of human existence, intricately linked to our very survival. At its core, the stress response is a complex interplay between various brain structures and neurochemical signals. Understanding this interplay allows us to not only recognize how stress affects our bodies and minds but also to harness this knowledge to manage stress more effectively.

Imagine you're facing a looming deadline at work. Almost immediately, your brain's intricate alarm system swings into action. The amygdala, often considered the brain's emotional center, processes threats and sends distress signals to the hypothalamus. This tiny yet crucial part of the brain acts as a command center, communicating with the rest of the body through the autonomic nervous system, priming you for a 'fight or flight' response.

The hypothalamus initiates the stress response by activating the sympathetic nervous system. Adrenal glands release adrenaline, which courses through your bloodstream in seconds. This cascade of activity is what leads to the typical symptoms of stress: increased heart rate, faster breathing, and heightened senses. These physiological changes are a primal mechanism, remnants of ancient times when such responses were crucial for survival against predators.

However, the stress response doesn't stop there. The hypothalamus also prompts the release of corticotropin-releasing hormone (CRH), which eventually leads to the secretion of cortisol from the adrenal cortex. Cortisol, often dubbed the "stress hormone," plays a significant role in maintaining fluid balance and blood pressure, providing the necessary energy in stress situations by allocating resources and diverting attention from less critical functions.

Normally, once the perceived threat has passed, cortisol levels drop, and the parasympathetic nervous system restores the body to a state of calm. But chronic stress disrupts this balance. Persistent stress keeps the hypothalamus in a prolonged state of alertness, leading to continuous cortisol production. This situation can lead to detrimental effects on various bodily systems, including the immune, digestive, and cardiovascular systems.

The brain's hippocampus and prefrontal cortex are somewhat like a checks-and-balances system for stress regulation. The hippocampus, which plays a central role in memory formation, reins in the hypothalamus, essentially telling it when to slow down the stress responses. Meanwhile, the prefrontal cortex, responsible for higher-level thinking and decision-making, helps assess whether a perceived threat is genuine. Chronic stress can damage these areas, reducing their ability to mitigate stress effectively and potentially leading to issues like anxiety and depression.

Yet, the relationship between stress and brain function is not all doom and gloom. Stress, in moderation, can actually enhance cognitive performance. This idea, described as 'eustress,' represents a beneficial form of stress that motivates and focuses the mind. The problem arises when stress becomes chronic and overwhelming, tipping the scale from helpful to harmful.

Interestingly, the brain's plasticity means it can adapt to stress over time. This adaptability, known as neuroplasticity, highlights the brain's ability to rewire itself in response to stress, learning, or injury. Positive coping mechanisms, such as mindfulness and exercise, can promote beneficial neuroplastic changes, enhancing one's ability to cope with stress and potentially reversing some negative effects.

The role of genetics should also be considered when examining stress responses. Genetic predispositions can influence how we respond to stress and the intensity of those responses. For example, certain gene variations affect the production and function of neurotransmitters like serotonin and dopamine, which can alter stress perception and response.

Despite genetic predispositions, environmental factors and personal experiences significantly shape one's resilience to stress. This notion highlights why people react differently to the same stressors. A person's upbringing, lifestyle, and even social support structures play pivotal roles in shaping their stress response mechanisms.

Furthermore, understanding the hormonal underpinnings of stress opens doors to new emotional regulation and stress management strategies. By leveraging specific practices that target brain structures involved in stress, one can develop more robust coping strategies. Techniques like mindfulness meditation, cognitive behavioral therapy (CBT), and biofeedback have been shown to calm the amygdala or enhance the prefrontal cortex's regulatory functions, reducing stress and its adverse effects.

In conclusion, stress is a multifaceted phenomenon deeply rooted in the brain's architecture and chemistry. Acknowledging the neuroscience behind our stress responses empowers us to manage stress proactively and effectively. By adopting healthy habits and developing emotional intelligence, individuals can leverage their understanding of stress biology to foster personal and professional growth. As we equip ourselves with knowledge, we unlock the potential to transform adversity into opportunity, driving forward not just our own well-being but that of those around us.

Techniques for Stress Management Using Brain Science

In today's fast-paced world, stress can feel like an inevitable part of modern life. Yet, understanding how our brains react to stress offers us a powerful tool for managing it effectively. By examining the neuroscience of stress, we can uncover reliable techniques that not only help manage stress but can also lead to personal transformation. This section delves into how we can harness scientific insights to restore balance and optimize our stress response.

One of the key mechanisms involved in stress management is the concept of neuroplasticity, which refers to the brain's remarkable ability to adapt and reorganize itself. Stress isn't just a response; it's a process that, with practice, can be molded. The brain's plasticity means we can train ourselves to manage stress more effectively by developing new neural pathways that favor adaptive thinking and coping behaviors. It's an exciting prospect that all of us can tap into with a bit of effort and patience.

Before diving into techniques, it's crucial to recognize how stress manifests in the brain. The primary player in the stress response is the hypothalamus-pituitary-adrenal (HPA) axis, a complex network that regulates reactions to stress and many physiological processes. This system releases cortisol, commonly known as the stress hormone. While short bursts of cortisol are manageable and can even be beneficial, chronic stress leads to sustained high levels that can impair brain function and overall health. Recognizing this is the first step to mitigating its effects.

Mindfulness meditation is perhaps one of the best-researched techniques for managing stress via brain science. Studies have shown it fosters changes in brain structures related to attention, emotion regulation, and mental flexibility. Regular mindfulness practice has been linked with reduced activity in the amygdala, the brain region responsible for the fight-or-flight response. In turn, this promotes a more measured response to stress triggers, enabling one to navigate stressful situations with greater calm.

Another transformative technique is the practice of cognitive reframing. This involves changing the way you perceive stressful situations, shifting from a threat-based mindset to one that views challenges as opportunities for growth. Through reframing, the prefrontal cortex, the part of the brain responsible for complex cognitive behavior and decision-making, is engaged. Over time, this trains the brain to create more resilient pathways, allowing for a more composed stress response.

Physical exercise also plays a crucial role in stress management, with profound effects on brain structure and function. Regular exercise increases the production of neurotrophic factors like BDNF (brain-derived neurotrophic factor), which supports the survival of existing neurons and encourages the growth of new neurons and synapses. Exercise can also lower levels of stress hormones and redirect energy in a more productive manner, leading to an enhanced mood and an improved ability to cope with stress.

Social connections and support networks can significantly mitigate stress by activating parts of the brain that deal with rewards and trust, such as the dorsal striatum and prefrontal cortex. Engaging in positive social interactions leads to the release of oxytocin, often dubbed the 'love hormone', which counteracts the effects of cortisol and lowers stress. Cultivating these bonds brings a twofold benefit: the mutual exchange of support not only strengthens relationships but also enhances psychological resilience.

Engaging in restorative activities like sleep is another powerful stress management technique. Quality sleep is vital for the consolidation of stress-reducing neural pathways. During sleep, the brain essentially resets, processing the day's stresses and enabling us to approach problems with renewed clarity and energy. Developing a consistent sleep schedule and creating a restful environment are foundational steps for harnessing the calming effects of slumber.

Additionally, understanding the role of nutrition in stress management shouldn't be overlooked. Nutrients such as omega-3 fatty acids, magnesium, and B vitamins have been shown to mediate stress responses by promoting healthy brain function. A balanced diet can thus contribute to stable moods and reduced reactivity to stressors, reinforcing the body's capacity to maintain equilibrium.

Lastly, techniques like biofeedback and neurofeedback offer innovative ways to gain insights into physiological stress responses and learn how to control them. These methods involve monitoring brain activity and physiological indicators in real-time, providing feedback that helps learn to self-regulate. By witnessing the direct impact of stress on the brain, individuals can cultivate greater self-awareness and develop tailored strategies to enhance personal stress-coping mechanisms.

Incorporating these diverse strategies rooted in brain science provides a comprehensive toolkit for addressing stress. They remind us that while stress is an inevitable part of life, we possess the agility and knowledge to navigate and even thrive under its pressure. Leveraging these insights not only aids in managing stress but also serves as a stepping-stone toward holistic emotional and cognitive growth. Armed with this understanding, we can reshape our stress responses, unlock potential, and foster a more harmonious existence.

Chapter 5:
Enhancing Self-Awareness

Stepping into the realm of self-awareness offers professionals and lifelong learners a powerful lens through which they can perceive both the self and the world more clearly. At its core, self-awareness invites a profound dialogue between the conscious and subconscious mind, bridging the gap through consistent reflection and introspection. Neuroscience shows that this internal conversation doesn't just heighten personal insight but activates specific neural pathways that connect past experiences with present emotions, thus fostering growth. Through practice and guided strategies, such as mindfulness and cognitive appraisal, individuals can transcend habitual behaviors and perceptions, unveiling a more authentic self. This journey not only enriches personal life but also enhances interpersonal relationships, creating ripples of emotional intelligence that extend far beyond the individual. Engaging actively with these elements positions one to leverage self-awareness as a robust tool for both personal and professional development, making the intangible aspects of growth become tangible and evident in everyday interactions and decisions.

Neurological Basis of Self-Reflection

In the vast landscape of the human mind, self-reflection acts as a compass guiding individuals towards deeper understanding and insight. But what enables this complex process at the neurological level? To delve into the mechanisms behind self-reflection, one must explore the intricate networks within the brain that make introspection possible. Notably, specific brain regions and their interactions play a pivotal role in facilitating the self-referential thoughts characteristic of self-reflection.

At the heart of self-reflection lies the default mode network (DMN), a network of brain regions that includes the medial prefrontal cortex (mPFC), posterior cingulate cortex (PCC), and the angular gyrus. These areas are highly active when your mind is at rest—when you think about your thoughts, desires, and emotions. This network essentially serves as the brain's internal dialogue forum, constantly engaged in considering who you are and how you relate to the world around you.

The medial prefrontal cortex, an integral part of the DMN, helps you ponder your identity. It's where the magic of self-awareness often begins. Whether you're contemplating a recent decision or revisiting past events, the mPFC allows you to process these experiences with a personal lens. But the mPFC doesn't work in isolation; it communicates with the PCC, which integrates complex information, weaving your past experiences with your internal state to produce a coherent self-narrative.

Across the brain's structure, interconnectivity is key. The interplay between the DMN and the limbic system—particularly structures like the amygdala—links the cognitive and emotional aspects of reflection. This connection enables an emotionally rich and contextually aware self-examination. The amygdala, known for its role in processing emotions, adds the emotional depth to self-reflection, allowing you to evaluate not only what you're thinking but also how those thoughts make you feel.

Now, consider the ventromedial prefrontal cortex, a subsection of the mPFC, which is critical for evaluating personal relevance. This region assesses which aspects of thought are meaningful or significant to you, acting as a filter that determines which reflections carry weight. It differentiates between fleeting thoughts and those that resonate deeply, thus impacting decision-making and personal growth.

The development of these brain regions and their functions isn't static. Neuroplasticity, the brain's ability to reorganize and form new connections, allows these networks to strengthen with practice. Engaging in regular self-reflection can reinforce these neural pathways, making introspection more intuitive. Imagine self-reflection as a mental gym session, where each session builds the mental muscles for deeper awareness.

Interestingly, studies show that self-reflection isn't only a byproduct of rest; it also contributes to emotional balance and resilience. Engaging with one's internal thoughts can lead to clarity and reduced stress. By better understanding the neural basis of reflection, one can appreciate how deliberate introspection cultivates resilience, providing solace during challenging times.

However, self-reflection isn't purely introspective. It has a profound impact on how individuals interact with the external world. By understanding one's desires, motivations, and emotional responses, people improve their interpersonal relationships and communication skills. This awareness isn't just advantageous on a personal level; it's crucial in professional settings, where emotional intelligence often correlates with leadership success and team synergy.

While much is known about these processes, the uniqueness of the human mind ensures that self-reflection remains a deeply personal experience. Cultural, social, and personal factors intertwine with neurological underpinnings to paint a reflective picture as unique as the individual exploring their mind. The blend of science and personal experience ensures that self-reflection remains as fascinating as it is invaluable.

As we move forward, exploring practical ways to enhance self-awareness will unlock further potential. In a world that often prioritizes external achievements, nurturing the internal conversation continues to build a foundation for authentic and holistic growth. While science provides the tools and understanding, it's the application that ushers transformation, paving the way for personal and professional evolution.

Strategies for Increasing Self-Awareness

In a world that's constantly in motion, it's often challenging to find a moment to pause and turn your focus inward. Yet, self-awareness—the ability to recognize and understand your own emotions, thoughts, and values—is a cornerstone of emotional intelligence. This quality not only enhances personal and professional relationships but also fosters self-discovery and growth. By leveraging findings from neuroscience, we can employ various strategies to boost self-awareness, creating a more profound understanding of ourselves.

One of the most effective strategies involves reflective journaling. By setting aside time each day to write about your experiences, thoughts, and emotions, you create a space for introspection. This practice encourages you to revisit your decisions and feelings, providing clarity over time. Neuroscience shows that the act of writing stimulates neural pathways, promoting deeper cognitive processing. As you journal regularly, patterns emerge, helping you identify triggers and emotional responses, bringing unconscious thoughts to consciousness.

Mindful meditation is another potent tool for increasing self-awareness. By focusing on the present moment without judgment, mindfulness meditation helps to quiet the noise in your mind and observe your internal state with clarity. Research indicates that regular meditation practice can alter brain structures, enhancing areas connected to attention and emotional regulation. As meditation deepens your sensory awareness, you become attuned to subtle shifts in your emotions and thoughts, granting you greater control over your reactions.

An often-overlooked strategy is soliciting feedback from others. While self-perception is invaluable, external perspectives can offer insights you might not otherwise consider. Constructive feedback from trusted colleagues, friends, or mentors can illuminate blind spots in your behavior and attitudes. Such feedback acts as a mirror, reflecting areas needing growth and reinforcing positive attributes. Neuroscience suggests that hearing and processing feedback could activate the brain's self-evaluation processes, enhancing adaptability and awareness.

Moreover, engaging in regular self-reflection, either informally or through structured techniques like SWOT analysis (Strengths, Weaknesses, Opportunities, Threats), can sharpen your self-awareness. Analyzing your strengths and weaknesses in various contexts provides insight into habitual patterns and thought processes. When you juxtapose these introspections with future opportunities and potential threats, it fosters a dynamic understanding of your internal landscape, equipping you with strategies for emotional adjustment and success.

The Socratic method of inquiry also plays a pivotal role in cultivating self-awareness. By asking probing, open-ended questions about your beliefs, values, and motivations, you challenge existing paradigms. This approach encourages critical thinking, compelling you to delve beneath the surface of automatic responses. As research suggests, engaging in deep questioning spurs brain pathways associated with problem-solving and creativity, enhancing your cognitive and emotional insight.

Additionally, mindful listening is crucial. Often, communication is seen as a dual exchange between speaker and listener; however, true dialogue involves listening with intention and openness. Mindful listening urges you to fully engage, processing not just the words but the emotions and intent behind them. Neuroscientists have found that attentive listening can build neural connections related to empathy and emotional recognition, leading to more nuanced self-awareness in social contexts.

It's essential to leverage modern technology as well, which can assist in self-monitoring and data collection for self-awareness enhancement. Apps and wearables can track mood fluctuations, stress levels, and even sleeping patterns, offering insights into your emotional states. These tools can serve as valuable adjuncts to personal reflection, giving empirical data that corroborates subjective insights. As technology evolves, its role in enhancing self-awareness grows, enabling real-time feedback and adaptive learning.

The path to increased self-awareness is neither linear nor straightforward; it requires patience and ongoing commitment. Neuroplasticity, the brain's ability to reorganize itself by forming new neural connections, assures us that change is possible at any stage of life. By consistently applying these strategies, you expand your capacity for self-discovery. As you become more attuned to your internal workings, you lay a solid foundation for emotional intelligence, empowering you to thrive in both personal and professional arenas.

In closing, increasing self-awareness is an evolving journey, enriched by strategies drawn from both ancient practices and contemporary neuroscience. Whether through reflective journaling, meditation, feedback, or technology, each strategy offers a different lens through which to explore your inner world. By embracing these methods, you commit to a path of growth, self-compassion, and empowerment, ultimately leading to a life of greater fulfillment and emotional mastery.

Chapter 6:
Building Emotional Resilience

In the realm of personal and professional growth, building emotional resilience acts as a powerful buffer against life's inevitable challenges. Harnessing the brain's remarkable ability to adapt, known as neuroplasticity, we can cultivate resilience by rewiring our neural pathways in response to adversity. This transformative process enables us to regulate emotional responses more effectively, fostering a mindset that not only withstands stress but thrives amidst it. By leveraging strategies rooted in neuroscience, such as cognitive reframing and mindfulness, we strengthen our emotional core, creating a robust foundation for navigating complexities in both our personal and work environments. This journey is less about avoiding difficulties and more about enhancing our capacity to recover from them, drawing on the brain's natural aptitude for growth and adaptation. Embracing the science of resilience, professionals and lifelong learners alike can move forward with renewed confidence and empowerment, equipped to face challenges head-on and harness them for deeper emotional intelligence.

Brain Mechanisms Behind Resilience

Resilience, the capacity to recover from setbacks and adapt effectively to challenges, is more than just a desirable trait; it's a complex neurological process that involves various brain structures working in harmony. At its core, resilience isn't merely a shield against adversity but a dynamic process powered by the brain's ability to adapt and reconfigure neural pathways. Understanding how our brain fosters resilience can empower us to enhance this quality, paving the way for greater emotional intelligence and growth in both personal and professional realms.

Central to this process is the prefrontal cortex, often heralded as the brain's control center. This region is profoundly implicated in planning, decision-making, and moderating social behavior. It enables us to control our impulses and develop long-term strategies rather than reacting instinctively. When we encounter stress or adversity, the prefrontal cortex helps us analyze the situation, consider various outcomes, and decide on a course of action. Its robust circuitry can strengthen resilience by enhancing our ability to stay calm, focus on solutions, and regulate our emotions effectively.

Yet, the prefrontal cortex doesn't work in isolation. It constantly interacts with the amygdala, a small almond-shaped structure responsible for processing emotions, particularly fear and pleasure. In potential danger, the amygdala automatically triggers a fight-or-flight response. However, a resilient brain manages this by allowing the prefrontal cortex to modulate the amygdala's response, ensuring we don't overreact but approach threats with calculated rationality. This balance allows us to experience fear not as a paralyzing force but as a motivator for thoughtful action.

Moreover, the hippocampus plays a pivotal role in resilience, especially in how we process and contextualize experiences. Situated in the limbic system, the hippocampus is integral to forming, organizing, and storing memories. Imagine facing a challenging event: the hippocampus tags these experiences as either positive or negative based on our emotional response and stores them for future use. A resilient brain can access and recontextualize these memories, learning from mistakes without being overwhelmed by past failures, turning adverse experiences into stepping stones for future success.

Resilience also hinges on neuroplasticity, the brain's remarkable ability to reorganize itself. This capability is not fixed but rather evolves with our experiences, emotional states, and even conscious efforts to change. Neuroplasticity enables the formation of new neural connections while pruning away less useful ones. This process solidifies our learning from overcoming challenges and helps us become more adaptable, allowing resilience to become ingrained in our neurological framework. The concept of "neurons that fire together, wire together" essentially captures this transformation, indicating that repeated resilient behaviors can forge robust mental pathways for dealing with future adversity.

Equally important is the role of neurotransmitters, particularly serotonin, dopamine, and oxytocin, in the resilience equation. Serotonin contributes to well-being and happiness, influencing mood balance and anxiety reduction. Regularly fostering environments or habits that boost serotonin levels—like social interactions, exercise, and nature exposure—can bolster resilience by elevating our ability to remain positive under stress. Dopamine, known for its role in reward processing, enables us to stay motivated and pursue goals despite setbacks, providing the neurological fuel necessary for resilience. Oxytocin, often dubbed the "love hormone," enhances bonding and trust, which are essential for drawing and lending emotional support during tough times.

However, identifying these brain mechanisms only scratches the surface. It's the dynamic interplay between these elements that truly defines resilience. When a person faces hardship, a well-coordinated response involving the prefrontal cortex's regulation, the amygdala's emotional incorporation, and hippocampal memory processing collaborates with neuroplasticity and neurotransmitter adjustments to support resilience. This integrated effort ensures that we bounce back, often stronger and more adaptable than before.

On a practical level, this understanding opens avenues to cultivate resilience actively. Practices like mindfulness and cognitive behavioral techniques can enhance neural pathways associated with resilience by encouraging the conscious regulation of thoughts and emotions. Mindfulness, in particular, induces physical changes in the brain such as increased prefrontal cortex activation, which can improve emotional regulation and stress responses, fortifying our mental resilience against adversity.

Furthermore, community support and nurturing relationships create an environment that neurologically supports resilience. Social interactions stimulate oxytocin release and engage the social brain, creating bonds that provide emotional fortitude. When we feel supported by others, our neurological response reduces stress hormone levels, further contributing to resilient outcomes. Here, professional growth dovetails with personal interaction, illustrating the comprehensive nature of resilience as both a cerebral and social construct.

In summary, the brain mechanisms behind resilience are as intricate as they are empowering. They reveal resilience not as a static trait but as an active capability that can be nurtured and expanded. This empowers individuals to confront challenges head-on, transforming adversities into opportunities for deeper self-awareness and emotional intelligence. As we navigate the complexities of everyday life and the professional world, understanding these biological bases provides valuable insights into how we can foster resilience, utilizing our brain's innate capabilities to thrive amidst uncertainty. Let this knowledge inspire both personal and communal growth, ushering in a brighter, more resilient future.

Cultivating Resilience through Neuroplasticity

In the domain of emotional resilience, the human brain unveils a fascinating capacity: neuroplasticity. This attribute empowers our brains to reconfigure and adapt in response to new experiences, learning, and even trauma. It is this adaptive quality that holds the key to cultivating resilience, a fundamental trait for both personal and professional growth. Neuroplasticity is not just about the brain's ability to change; it's about transformation and possibility, a concept that encourages us to seek out and embrace change as an opportunity for growth.

At its core, neuroplasticity involves the brain's ability to reorganize itself by forming new neural connections. This process isn't limited by age or circumstance; it can occur throughout our lives, paving the way for continuous emotional development. Our experiences, thoughts, and actions influence the brain's wiring, meaning we are not entirely defined by our pasts. Instead, we can actively mold our emotional responses, attitudes, and perceptions, an empowering notion that suggests we hold substantial influence over our futures.

Imagine neuroplasticity as the brain's innate ability to rebound. Just as a rubber band stretches yet returns to its original form, our brains can recover and even thrive from adversity. The elasticity of our neurological pathways is vital to developing resilience, as it enables us to cope with stress and adapt to new challenges effectively. This adaptability is essential for maneuvering the complexities of modern life, especially in environments that demand a high level of emotional intelligence.

Research has shown that individuals can harness neuroplasticity to build resilience through targeted practices. Techniques such as mindfulness, meditation, and positive visualization are instrumental in fostering neurological growth and resilience. These methods work by promoting a state of mental focus and clarity, reducing stress, and enhancing emotional regulation—all of which are critical elements in rewriting the brain's response to challenging experiences.

Within a work setting, neuroplasticity facilitates learning and adaptation, allowing professionals to tackle novel problems creatively and effectively. This ability is crucial in fast-paced industries where change is the only constant. By consciously engaging in activities that promote neuroplasticity, such as continuous learning and reflective practices, individuals not only improve their professional skills but also bolster their capacity for emotional resilience.

The path to cultivating resilience through neuroplasticity isn't a solitary journey. Social interactions and supportive relationships can significantly influence the brain's adaptive processes. Being part of a community provides emotional support and introduces diverse perspectives, which can be invaluable when navigating life's hurdles. These interactions help reinforce neural pathways associated with empathy, compassion, and cooperative problem-solving, further underlining the intricate connection between social connections and emotional resilience.

Inspiration for enhancing neuroplasticity can also be drawn from understanding the brain's response to positive experiences. Engaging in activities that elicit joy and satisfaction stimulates the release of neurochemicals like dopamine and serotonin. These chemicals not only boost mood but also support the rewiring of neural pathways, fostering an environment where resilience can flourish. Encouraging a positive outlook and practicing gratitude are simple yet powerful strategies that can transform our mental landscape.

The journey towards resilience is intertwined with self-awareness. Recognizing our cognitive and emotional patterns is the first step in leveraging neuroplasticity. Reflective practices, such as journaling or therapy, enable us to map out these patterns and understand the triggers that affect our emotional responses. With greater self-awareness, we can consciously choose to redirect our thoughts and reactions, strengthening more adaptive neural connections over time.

It's important to acknowledge that cultivating resilience through neuroplasticity is an ongoing process. Just as our environments and experiences evolve, so too should our strategies for nurturing resilience. This requires a commitment to lifelong learning and the willingness to adapt and integrate new approaches. Embracing this mindset not only enhances our emotional resilience but also enriches our lives, empowering us to navigate the ebbs and flows of human experience with grace and strength.

Finally, the integration of technology offers novel avenues for amplifying neuroplasticity-driven resilience. From apps that facilitate mindfulness practices to virtual reality environments that offer stress-relief simulations, technology can serve as a powerful ally in reshaping our brains. However, as with any tool, the key lies in its mindful application—ensuring that technology enhances, rather than diminishes, our emotional landscapes.

Embracing resilience through neuroplasticity is a transformative journey, one where each step helps unravel the potential that resides within the patterns of the brain. It affirms the idea that our emotional capacity isn't static but dynamic and ever-expanding. As professionals and lifelong learners, tapping into this profound power isn't just about overcoming challenges, it's about thriving amidst them, unlocking both personal and collective human potential.

Chapter 7:
Empathy and the Brain

Empathy isn't just an emotional experience; it's a complex neurological dance that plays a critical role in our everyday interactions and emotional growth. At the heart of empathy lies a fascinating interplay between various neural circuits, involving regions of the brain like the anterior insula and the anterior cingulate cortex, which enable us to feel others' emotions as though they were our own. This connection forms the bedrock of our ability to understand and relate to one another genuinely. By harnessing our brain's remarkable capacity for empathy, we can transform our personal and professional relationships, unlocking richer, more meaningful connections. Science-backed techniques, rooted in neuroscience, allow us to enhance these empathic pathways, boosting not only emotional intelligence but also our interpersonal skills. Engaging with these practices invites us to look beyond ourselves, encouraging a more profound comprehension of the shared human experience and nurturing a world where understanding prevails over discord.

Neural Circuits of Empathy

Empathy, at its core, is the ability to experience and understand the emotions of others. It's a defining trait of our humanity, allowing us to connect with, comfort, and support those around us. To unravel the intricate mechanisms that drive empathy, we need to delve into the brain's neural circuits. At a neurological level, empathy is not a singular process but a complex network involving multiple brain regions each weaving its own thread into the tapestry of empathetic understanding.

One of the key players in this network is the *anterior insula*. It's a small region hidden within the lateral sulcus and is pivotal for processing bodily emotions. When you see someone experiencing an emotion, your anterior insula activates as if you're going through the same emotions yourself. This mirroring mechanism is critical for the empathetic process because it allows us to 'catch' the emotional experience of others. This is often why we physically cringe when watching someone in pain or smile when we witness joy—it's a shared emotional journey dictated by this part of the brain.

The *amygdala*, often known as the brain's emotion center, is another crucial component. While its role in fear and aggression is well-documented, it also processes the emotional significance of social stimuli. It deciphers the emotional states conveyed through facial expressions and body language, providing quick assessments that help guide our social interactions. These amygdalar assessments are automatic, influencing our unconscious biases and reactions to others' emotional cues, highlighting its essential role in both empathy and social behavior.

Adjacent to the amygdala, the *ventromedial prefrontal cortex (vmPFC)* plays a contributory role in empathy by supporting affective processing and emotional regulation. It integrates emotional responses with higher-order cognitive processes, facilitating decisions that incorporate empathy. The vmPFC is responsible for an individual's ability to imagine the emotional experiences of others and to make prosocial emotions such as compassion prevail over potentially aggressive responses. Thus, it acts as a mediator between emotional reactions and rational thoughts, shaping how empathetical behaviors manifest.

Another critical section of this network is the *temporoparietal junction (TPJ)*. It's essential for perspective-taking and differentiating one's own intentions from those of others. The TPJ allows us to step into another person's shoes, walking that proverbial mile and understanding their mental and emotional states. This mentalization capability is not only foundational to empathy but also pivotal to collaborative and harmonious social interaction, which results in better personal and professional relationships.

Furthermore, the brain's mirror neuron system deserves attention in this discussion. Discovered in the premotor cortex and extending its influence into the parietal lobes, mirror neurons activate both when an individual performs an action and when they observe the same action performed by someone else. This mirroring is a neurological underpinning of our capacity to understand another's actions and intentions, serving empathy by allowing us to relate through shared experiences. In this way, mirror neurons enable us to 'feel with' others, bridging the gap between personal experience and empathetic understanding.

The intricate interplay of these brain regions illustrates that empathy is not merely an abstract concept but a well-coordinated symphony played by our neural circuits. Developing a better understanding of these circuits opens up numerous avenues for enhancing empathy in individuals through targeted interventions and training.

Given the dynamic nature of the brain, research suggests that empathy can be cultivated and strengthened through specific practices designed to engage and enhance these neural pathways. For example, mindfulness techniques can increase activity in areas responsible for empathy and reduce emotional distress stemming from emotional overload. By regulating the amygdala's responsiveness, such practices can minimize knee-jerk reactions and foster more intentional empathetic responses.

The implications of nurturing empathy are profound, ranging from individual well-being to societal improvements. Empathy fosters communication and mutual respect, and it's foundational to effective leadership and collaboration. Consequently, organizations increasingly recognize the value of empathy in enhancing workplace relationships and boosting overall morale and productivity.

Still, there are challenges. Not every individual's brain reacts the same way due to variations in neural wiring, past experiences, and even genetic factors. However, with continued research and practice, it remains entirely possible to nurture the parts of the brain engaged in empathetic processes. Cognitive-behavioral strategies, empathy training, and perspective-taking exercises can all contribute to more empathetic behavior.

It's essential to highlight the dichotomy the neural circuits of empathy represent. On one hand, they enable deep social connections, fostering a shared sense of humanity. On the other, they can differentiate us from those we perceive as 'other,' underscoring the need for conscious effort to build bridges across diverse backgrounds and experiences. Recognizing this, the conscious practice of empathy becomes a tool for both personal growth and broader societal cohesion.

Empathy is not just an innate trait we possess but a skill we can develop. By understanding the brain regions at play, we become more equipped to consciously apply empathy in daily interactions. As we harness these neural insights, we create a ripple effect: enhancing interpersonal relationships, improving professional environments, and contributing to a more compassionate world. Transforming these scientific findings into practical applications is the next frontier in using neuroscience for expansive emotional and societal growth.

Empathy and its root in our brain's architecture paint a motivational picture: our connection to others is hardwired yet flexible, inherent but enhanced by choice. This knowledge challenges us to actively engage with the aspects of our brain that serve empathetic growth. Undoubtedly, the potential of empathy is as vast as the neural networks that facilitate it, urging us to explore and embrace empathy's power for the betterment of humanity.

Developing Empathy through Neuroscientific Techniques

In the heart of our interconnected world lies empathy—a crucial compass guiding our interactions with others. It's more than a mere social nicety; empathy's deeply embedded in our cerebral architecture. Understanding this complex human ability requires not just an exploration of the neural circuits of empathy but also a strategic approach to nurture and expand it. Neuroscience offers a treasure trove of techniques to develop empathy, leveraging the flexibility of our brain's own wiring, also known as neuroplasticity.

First and foremost, let's delve deeper into the role of the mirror neuron system. These neurons are often referred to as the brain's "social network." They fire not only when we perform an action but also when we observe others performing the same action. This phenomenon allows us to inhabit others' perspectives, offering a window into their emotional experiences. When harnessed consciously, this understanding can form the bedrock of enhanced empathetic connections, transforming our reactive instincts into thoughtful responses.

Integrating mindfulness practices can drastically bolster our empathic abilities too. Mindfulness, the art of being intensely aware of the present, without judgment, can refine our attention and emotional resonance with others. Regular mindfulness exercises enhance the connectivity between the amygdala and prefrontal cortex, regions responsible for emotional regulation and decision-making. This improved neural pathway facilitates a calmer, more grounded engagement with the emotional states of others, allowing for deeper empathy.

Empathy training programs rooted in neuroscientific principles also present promising avenues for personal and professional growth. These programs often emphasize compassionate observation, an upgraded version of active listening. By refining our listening skills through training techniques focused on sympathetic engagement and perceptual sensitivity, we can genuinely understand and, more importantly, feel what others are experiencing. Over time, these practices can lead to substantial brain structural changes, making empathy a default response rather than a conscious choice.

Another key neuroscientific technique involves perspective-taking and mentalizing. This involves imagining the world from another person's viewpoint, consciously stepping into their shoes. This practice is backed by neuroscience, showing findings that the medial prefrontal cortex is significantly activated during perspective-taking tasks. Repeated engagement in such exercises can enhance the brain's ability to process and respond to emotional cues, framing empathy as a skill that can be learned and finetuned over time.

Furthermore, immersive virtual environments are emerging as state-of-the-art tools to cultivate empathy. These technologies enable users to experience scenarios from another person's perspective, promoting understanding through real-time emotional engagement. Such experiences stimulate the brain's empathic neural circuits, fostering an emotional evolution that can be transferred to real-world interactions.

In terms of practical application, empathizing through storytelling is another effective technique. Stories have a profound impact on the brain, activating the neural regions associated with both language comprehension and empathy. This enhances our ability to build connections with others. By engaging with diverse narratives, we stimulate the brain's emotional response systems, reinforcing empathy as we actively broaden our understanding of different life experiences.

Emotional mapping is a lesser-known but equally powerful technique. By charting out our own emotional responses and patterns, we become aware of how our emotions influence our interactions. This introspective practice supports the development of empathy by highlighting the interconnectedness of our emotions and those of the people around us. The brain's lateral prefrontal cortex plays a primary role in this analytical process, reinforcing our capacity for understanding nuanced emotional dynamics.

It is also important to note the role of social play in empathy development. Engaging in cooperative play activities can enhance our ability to read social cues and foster empathetic interactions. These activities stimulate brain areas linked to social processing and cooperation, making social play not just a method of entertainment but a crucial practice for empathy building. Such engagements encourage synchrony between the emotional and social cognition networks in the brain, reinforcing our empathic aptitude.

Advancing empathy is not a solitary endeavor but rather a collaborative journey with profound biological roots and contemporary applications. Neuroscience equips us with the tools to consciously enhance this ability, framing empathy not only as a result of intrinsic structures but as a dynamic skill that we can cultivate and expand throughout our lives. By utilizing these neuroscientifically-backed techniques, we place ourselves on a path to becoming not only more effective communicators but also more compassionate and understanding individuals fuelled by the brain's remarkable capacity for growth and adaptation.

Chapter 8:
Emotional Regulation Techniques

Emotional regulation is a vital skill for personal and professional growth, intertwined with our brain's fascinating neurocircuitry. Grounded in scientific understanding, the brain's role in emotion spans from the amygdala's initial reactions to the prefrontal cortex's more nuanced judgments. Cultivating robust emotional regulation empowers you to navigate life's highs and lows with grace and agility. This chapter delves into the harmonious dance of neural connections and offers practical strategies to enhance emotional control, allowing you to harness your emotions as allies rather than adversaries. By embracing these techniques, you're not just altering immediate emotional responses but also fostering long-term neurological resilience. Inspired by your journey toward growth, tap into the power of mindfulness, cognitive restructuring, and neuroplasticity, paving the way for a refined emotional intelligence that elevates both interpersonal interactions and inner peace. It's about mastering the mind for a more balanced, fulfilling life.

Brain's Involvement in Emotional Control

The human brain is an extraordinary organ, with a significant influence on our emotional lives. At its essence, emotional control is a highly orchestrated process, deeply embedded within our neural framework. Imagine the brain as a complex symphony, where different sections collaborate to produce the music of our emotions. The prefrontal cortex, amygdala, and hippocampus are the virtuosos of this ensemble, each playing an essential role in regulating emotions.

Let's delve into how these brain structures interact to influence emotional control. The prefrontal cortex, often associated with complex cognitive behavior, personality expression, and decision-making, acts as the conductor. It supervises and adjusts our emotional responses, ensuring they suit the context of our surroundings. It enables us to pause, evaluate a situation, and choose a suitable response, rather than reacting impulsively.

The amygdala, a small almond-shaped cluster of nuclei, is integral in processing and eliciting emotional responses, especially those related to fear and pleasure. Think of it as the emotional sentinel, efficiently scanning the world for potential threats and regulating our fight-or-flight responses. When perceived dangers arise, the amygdala sounds the alarm, prompting the prefrontal cortex to evaluate the situation further.

The hippocampus, another key player, is intertwined with memory formation and retrieval. It provides context to our emotional experiences, connecting past memories with present circumstances. By remembering the past, the hippocampus informs the amygdala and prefrontal cortex of historical precedents, subtly influencing our emotional reactions to familiar contexts.

Research has shown that these structures don't work in isolation; rather, they are part of a complex network. Functional MRI scans have revealed the dynamic interplay between these areas during emotional regulation tasks. Understanding this interplay not only sheds light on how the brain manages emotions but also highlights areas where intervention can enhance emotional intelligence.

In everyday life, the prefrontal cortex's ability to regulate the amygdala's response plays a pivotal role in managing stress, inhibiting inappropriate reactions, and fostering resilience. This control helps us maintain composure under pressure and react with empathy rather than anger in challenging situations. Hence, individuals with a well-developed prefrontal cortex are often better at navigating social environments and maintaining emotional balance.

Yet, this system isn't foolproof. When overwhelmed by stress or trauma, the amygdala can hijack rational thought, overriding the prefrontal cortex. Under such circumstances, emotions can become unmanageable, leading to anxiety, aggression, or panic. These instances illustrate the importance of maintaining brain health and developing strategies to enhance emotional regulation capacities.

Fortunately, neuroplasticity offers a ray of hope. The brain's remarkable ability to reorganize itself by forming new neural connections provides a foundation for improving emotional regulation. Mindfulness practices, cognitive behavioral therapies, and even regular physical activity can strengthen the neural circuits responsible for emotional control. These practices enable individuals to train their brains, much like a muscle, enhancing the prefrontal cortex's regulatory abilities.

Moreover, consciously engaging in activities that require emotional regulation can bolster these neural pathways. Whether it's through meditation, reflective journaling, or discussions that require calmness and empathy, each practice contributes to forging robust emotional control mechanisms. Over time, with consistent practice, these strategies become ingrained, leading to better mental flexibility and emotional agility.

Developing emotional control isn't merely about suppressing negative emotions. It's about understanding and channeling them effectively. By acknowledging that emotions carry information about our needs and desires, one can use them as a guide rather than a hindrance. This perspective shift allows emotional control to become a tool for personal growth and interpersonal success.

In professional settings, individuals with heightened emotional control often excel in leadership roles. Their ability to manage emotional responses ensures clear communication and thoughtful decision-making, fostering a climate of trust and collaboration. In personal contexts, such control nurtures healthier relationships, as it reduces misunderstandings and enhances empathetic engagement.

Integrating neuroscience insights into emotional regulation can empower us to fine-tune our emotional responses. This fusion of science and self-awareness encourages a future where emotional mastery isn't just accessible but achievable. As we deepen our understanding of the brain's role in emotional control, we can cultivate an emotionally intelligent society that thrives on empathy, balance, and resilience.

Tools and Strategies for Emotional Regulation

In the quest to attain emotional mastery, the arsenal of tools and strategies for emotional regulation emerges as a significant ally. These techniques involve harnessing the complex workings of our brain to manage, modulate, and channel emotions effectively. A well-regulated emotional state is not only pivotal for personal well-being but is also essential in professional settings, where interactions are frequent and often intense. Understanding that emotions can be deliberate responses rather than spontaneous reactions forms the first step in mastering emotional regulation.

At the core of emotional regulation strategies lies the belief that our thoughts significantly influence emotions. Cognitive reappraisal, or the art of reinterpreting a situation to alter its emotional impact, stands out as a prominent technique. For instance, perceiving a high-pressure work presentation not as a threat but as an opportunity can transform anxiety into excitement. By refining our thought processes, we essentially reshape our emotional responses, thereby gaining control over our emotions rather than being at their mercy.

Another tool that has gained traction is deep breathing exercises. While often underestimated, controlled breathing can directly reduce the activity of the amygdala, the brain's emotional alarm system. Slow, deliberate breaths help shift the nervous system from a stress-fueled state to one of calmness, allowing clarity in emotional storms. When faced with overwhelming feelings, the simplicity of just a few deep breaths can offer respite and grounding in tumultuous moments.

Equally vital is developing the habit of mindfulness. Mindfulness involves paying attention to the present moment without judgment and with heightened awareness. This practice not only fosters greater self-awareness but also helps in identifying and understanding emotional triggers. By being present, we can observe emotions as transient experiences, which provides the space to choose how to respond or even if to respond at all. Consistent mindfulness practice can restructure neural pathways, encouraging a more balanced and less reactive emotional baseline.

On the journey to emotional regulation, cultivating an environment conducive to positive emotional states can be highly beneficial. Our surroundings, social interactions, and daily routines can either fuel emotional dysregulation or contribute to emotional harmony. Regular exposure to positive influences—such as supportive relationships, soothing music, or even nature—can bolster our emotional resilience and regulation abilities.

Emotion regulation isn't just about managing negative emotions; it also involves amplifying positive ones. Techniques such as savoring, where one focuses on the joy of positive experiences, can enhance well-being. Similarly, the deliberate practice of gratitude can bring about a profound shift in emotional states, helping individuals focus on what's going well rather than dwelling on challenges. By regularly tuning into positive experiences and maintaining a gratitude practice, we can foster a more optimistic emotional climate.

Social support systems play a crucial role in emotion regulation. Engaging with empathetic friends, mentors, or even professional therapists can provide perspectives and tools that might not be apparent when navigating emotions solo. The act of verbalizing feelings not only provides relief but often leads to insights necessary for emotional growth. Dialogue and empathetic exchanges produce hormonal responses that reinforce emotional stability and resilience.

While these strategies vary from one individual to another, the overarching theme remains—intentionality. An intentional effort to use these tools consistently equips us with the ability to regulate emotions more effectively over time. As our brain commits these practices to muscle memory, emotional regulation becomes less a task and more an effortless habitual response. This transformation leverages the brain's neuroplasticity, slowly but surely altering the neural circuits related to emotion regulation.

The journey of emotional regulation is highly personal, yet universally beneficial. By adopting a mix of cognitive and behavior-based strategies, individuals equip themselves not only to manage their emotional lives but enhance their overall functioning, personal relationships, and professional performance. As you explore these tools and make them your own, remember that like any skill, emotional regulation improves with practice and perseverance. Yet, it's in this effort and commitment that the seeds of profound personal and professional growth are sown.

Chapter 9:
The Role of Neurochemicals in Emotion

Exploring the intricate world of neurochemicals reveals much about how our emotions take shape, function, and influence our daily lives. These powerful messengers—like dopamine, serotonin, and oxytocin—act as the brain's architects of mood and feeling, each playing a distinct yet interconnected role in weaving the fabric of our emotional experiences. While neurotransmitters ignite swift communication across neurons, hormones gently modulate longer-term mood states. Balancing these chemical levels is crucial for emotional well-being, opening up opportunities for deeper emotional understanding and better management of our emotional states. By understanding and adjusting these neurochemical pathways, we empower ourselves to navigate emotional landscapes with greater agility and resilience, harnessing the brain's potential to foster growth, harmony, and fulfillment in both personal and professional realms. With this knowledge, you're not only comprehending emotions on a biological level but also leveraging that understanding to enhance your emotional intelligence, paving the way for more profound interpersonal connections and enriched life experiences.

Understanding Neurotransmitters and Hormones

At the core of our emotional experiences are complex biochemical messengers known as neurotransmitters and hormones. These compounds flow through our neural networks and bloodstreams, shaping not only how we feel but how we interpret and react to the world around us. By understanding the role these chemicals play, we can begin to harness them for emotional mastery, enhancing both personal growth and professional development.

Neurotransmitters are chemical messengers that primarily function within the nervous system. They relay signals across synapses, the gaps between neurons, allowing for the transmission of thoughts, emotions, and physical sensations. Serotonin, dopamine, and norepinephrine are among the most studied neurotransmitters due to their significant impact on mood regulation and emotional well-being.

Serotonin, often dubbed the "feel-good" neurotransmitter, plays a crucial role in mood stabilization. Low levels of serotonin are commonly associated with depression and anxiety, whereas higher levels tend to promote feelings of happiness and contentment. Understanding how serotonin functions can be empowering. For instance, lifestyle changes such as regular exercise and exposure to natural sunlight are known to boost serotonin levels, offering a natural avenue to enhance our mood.

Dopamine, the neurotransmitter often linked with the brain's reward system, governs our experience of pleasure and motivation. When dopamine levels are optimal, we feel energized and engaged. However, imbalances can lead to decreased motivation and a lack of pleasure, evident in conditions such as depression and addiction. It's a reminder of the intricate balance our brains maintain and how pivotal that balance is for our productivity and satisfaction.

Norepinephrine serves as both a neurotransmitter and a hormone, underscoring its critical dual role. Known chiefly as a neurotransmitter involved in arousal and alertness, norepinephrine is essential for the fight-or-flight response, preparing the body to face challenges or stressful situations. Its influence on attention and responding actions makes it a key player in our emotional and cognitive responses.

Beyond neurotransmitters, hormones act as messengers in the bloodstream, regulating long-term processes like growth, metabolism, and reproduction. Cortisol, often called the "stress hormone," highlights the dual nature of hormones in emotional processes. While cortisol is crucial for managing stress, helping the body respond effectively to short-term challenges, chronic stress leads to prolonged elevated cortisol levels, damaging health and impairing cognitive functions.

Oxytocin, another hormone, has often been celebrated as the "love hormone." It's released in response to social bonding and physical touch, promoting feelings of trust and empathy. Understanding oxytocin's role emphasizes the importance of fostering strong social connections and nurturing relationships to enhance emotional well-being.

The interaction between neurotransmitters and hormones is a dance of balance and interconnectedness. This balance is not static but dynamic, influenced by genetics, environment, lifestyle choices, and psychological states. This knowledge shines a light on the pathways through which we might cultivate emotional intelligence, offering a framework for intentional action.

Enhancing emotional intelligence involves more than just understanding these biochemical foundations; it's about leveraging this understanding to drive personal transformation. Techniques such as mindfulness, exercise, and social engagement can optimize neurotransmitter and hormone levels, essentially tuning the brain's emotional orchestra.

For professionals aiming to enhance their emotional and interpersonal skills, appreciating the nuances of these chemical messengers is indispensable. By learning to recognize how these compounds influence mood, behavior, and decision-making, one can develop strategies to manage emotions effectively, improve relationships, and lead more fulfilling lives.

Embracing the science of neurotransmitters and hormones isn't just about academic knowledge—it's about transformation. By consciously taking steps to balance these chemicals, we align our internal states with our external goals, catalyzing growth in every domain of life. Whether in leadership, teamwork, or personal interactions, the awareness and mastery of our emotional biochemistry empower us to thrive.

Balancing Neurochemicals for Emotional Well-being

In the intricate dance of emotions, neurochemicals orchestrate the rhythms that we call feelings. Understanding how these substances influence our emotions is just the first step. The real challenge lies in learning how to balance them to nurture emotional well-being. This balance is no simple feat, as various neurotransmitters and hormones contribute differently to our emotional landscape. Yet, by understanding and adjusting lifestyle choices, individuals can significantly impact their neurochemical states for the better.

Serotonin, often labeled the "happiness hormone," plays a critical role in mood stabilization. It's produced primarily in the gut and influences a variety of psychological processes. Higher levels of serotonin can lead to feelings of happiness and calm, while deficits are linked with depression and anxiety. One effective way to boost serotonin levels is through exposure to sunlight. Just as flowers track the sun, our bodies thrive when basked in natural light, triggering biochemical processes that enhance serotonin production. Engaging in regular outdoor activities not only provides this necessary exposure but also adds a dimension of physical exercise that further promotes well-being.

Dopamine, another vital neurotransmitter, is often associated with pleasure and reward. It drives behavior by reinforcing actions that bring joy or satisfaction. The cycle of anticipation and reward creates a dopamine loop, crucial for motivation. However, an imbalance—particularly excess—can lead to addictive behaviors, as the brain relentlessly seeks that next "hit" of pleasure. Balancing dopamine isn't about abstaining from rewarding activities but instead enjoying them in moderation. Practicing mindfulness can help individuals identify and appreciate these rewards without becoming dependent on them, fostering healthier dopamine regulation.

Norepinephrine, sometimes referred to as noradrenaline, acts as both a neurotransmitter and a hormone. It's integral to our body's fight or flight response, heightening alertness and focus in times of stress. While attributable to life's ebb and flow, chronic stress can lead norepinephrine levels to remain elevated, contributing to anxiety and tension. Balancing this chemical requires intentional stress management techniques. Practices such as yoga, meditation, and deep breathing are proven methods to lower norepinephrine levels, promoting relaxation and a sense of peace amid chaos.

Oxytocin, often called the "love hormone," is pivotal in building and maintaining social bonds. It's released during moments of meaningful connection, such as hugging or nurturing relationships. This powerful neurochemical enhances feelings of trust and reduces fear. Interestingly, oxytocin's effects stretch beyond interpersonal interactions, influencing stress responses as well. Enhancing oxytocin levels can be as simple as fostering deeper connections with others or volunteering, activities that naturally lead to more opportunities for meaningful, emotion-rich exchanges.

Understanding the interplay between these neurochemicals allows individuals to cultivate a more refined emotional balance. Consider the role of diet in this process. What we consume profoundly affects neurotransmitter production. Foods rich in omega-3 fatty acids, such as fish and flax seeds, support serotonin and dopamine function, while antioxidant-rich fruit and vegetables combat oxidative stress, protecting neurotransmitter pathways. It's a symbiotic relationship: balanced neurochemicals promote healthy choices, and healthy choices, in turn, reinforce neurochemical stability.

Exercise is another cornerstone in managing and balancing neurochemicals effectively. Physical activity prompts the release of endorphins, the body's natural painkillers, and mood elevators. Simultaneously, it influences serotonin and dopamine production, enhancing mood and motivation. Regular aerobic exercise is particularly effective, providing a natural and holistic avenue to release pent-up stress and boost happiness levels, making it a vital part of any balanced emotional health regimen.

Sleep cannot be overstated in its importance for neurochemical equilibrium. During restful slumber, the brain repairs and recalibrates, crucially regulating hormone and neurotransmitter levels. Poor sleep disrupts this balance, exacerbating stress and impairing cognitive functions such as decision-making and emotional regulation. Establishing a regular sleep schedule and creating a restful night routine are practical tools to ensure that the body's neurochemical systems have the chance to reset and restore.

While individual efforts are crucial, social environments also play a significant role in neurochemical regulation. Supportive social networks can buffer against stress and depression, enhancing neurochemical health. Engaging with a community creates opportunities for oxytocin release and cultivates a sense of belonging and safety, crucial for maintaining mental well-being and resilience.

Ultimately, achieving neurochemical balance doesn't mean eradicating negative emotions but rather learning to navigate them skillfully. By consciously adjusting lifestyle factors like diet, exercise, sleep, and social interaction, individuals can manage their neurochemical states effectively. This proactive approach allows them to maintain a steady emotional keel, navigating life's ups and downs with resilience and grace. The journey toward emotional well-being is ongoing, but with awareness and effort, a more harmonious state of being is well within reach.

Chapter 10:
Social Connection and
Brain Health

Our brains are wired for social connection, a fundamental aspect of human existence with profound implications for brain health and emotional mastery. Engaging in meaningful social interactions can stimulate the release of key neurochemicals that promote happiness and reduce stress. These interactions foster a sense of belonging and can even enhance cognitive function, showing that our connections are not just pleasant; they're essential. Strengthening social bonds doesn't just amplify our emotional intelligence but also creates a supportive network that can aid personal growth and resilience. Understanding the neural underpinnings of social interaction reveals paths to nurture these connections purposefully, ensuring we harness their full potential for personal and professional development. By prioritizing social engagement and understanding its neurological benefits, we can foster an enriching environment that encourages continuous emotional growth and nurtures a healthier brain.

Neural Underpinnings of Social Interaction

The human brain is a marvel of complexity, intricately wired to foster social relationships that shape our lives. Social interaction, a core aspect of being human, is not a mere byproduct of our behavior but a fundamental component engrained within our neural architecture. Understanding the brain's role in social connection opens a pathway to harnessing these interactions for emotional and cognitive growth, enhancing our overall well-being and effectiveness.

At the heart of social interaction lies the prefrontal cortex, a region of the brain crucial for decision-making, personality expression, and moderating social behavior. It's where we process social cues and navigate the intricate mesh of social dynamics. This region helps us read others' emotions through facial expressions, body language, and vocal tones, allowing us to respond appropriately. Our neural circuits are fine-tuned to these subtle signals, processing them with remarkable speed and accuracy.

Additionally, the limbic system, including the amygdala and the anterior cingulate cortex, plays a pivotal role in our emotional responses to social stimuli. The amygdala is particularly sensitive to fear and anger, triggering responses that can range from stress to empathy. When you meet someone for the first time, it is this part of the brain that helps you gauge trustworthiness and determine emotional intentions. The anterior cingulate cortex mediates pain and emotional regulation, a key player in understanding the social pain of exclusion or rejection.

Empathy, another critical component of social interaction, is rooted in the brain's mirror neuron system. These neurons don't just enable us to understand others' actions but allow us to feel their emotions as if they were our own. This neural mirroring creates a shared emotional space, essential for forming deep connections. By understanding someone's joy or suffering, we build rapport and contribute to emotional intelligence. This action-perception cycle forms the basis for responding empathetically and fostering meaningful social bonds.

Moreover, neurotransmitters like oxytocin and serotonin significantly influence our social behaviors. Oxytocin, often dubbed the "love hormone," enhances bonding and trust between individuals. It's released during positive social interactions, such as physical touch or shared laughter, reinforcing these experiences. Serotonin impacts mood and social behavior, contributing to the sense of well-being and community belonging. Balancing these neurochemicals can profoundly impact how we interact socially.

Complex social networks require our brains to be adaptive, as the rules of engagement can change rapidly. Here lies the importance of neuroplasticity—the brain's ability to reorganize itself by forming new neural connections. Social skills are not static; they're enhanced through experience and practice. Like a muscle, the brain strengthens these connections through consistent interaction, confirming that social competence can be developed with intention and effort.

The impacts of social interaction extend beyond emotional benefits. Engaging positively with others can significantly enhance cognitive functioning. Socially active individuals are often more creative and show improved information processing abilities. Conversations encourage the brain to form new ideas and perspectives, inspiring creativity as a result of diverse interaction.

Furthermore, social interactions can serve as a buffer against neurodegenerative diseases. Engaging in meaningful conversations, maintaining friendships, and feeling part of a community can mitigate cognitive decline. Loneliness, conversely, has been associated with increased risks of dementia and other health issues, underlining the importance of nurturing social connections for brain health across the lifespan.

The integration of social neuroscience into everyday life offers powerful tools for personal development. By consciously enhancing our social environments, we can influence the neural structures that lead to better emotional management and mental health. This involves not just connecting with others but doing so mindfully, with an awareness of how these interactions are shaping our brain and emotional states.

Techniques such as active listening, empathetic communication, and emotional validation can reinforce positive neural pathways. They encourage the use of the prefrontal cortex for more effective responses and reduce impulsive reactions that often emanate from the amygdala. Additionally, these techniques can be practiced and honed, leveraging the brain's plastic nature to develop superior social skills.

Ultimately, understanding the neural underpinnings of social interaction empowers us to take control of our relational dynamics. Whether in personal life or professional environments, the ability to navigate social landscapes with empathy and intelligence enhances not just our interactions but also our mental resilience. By focusing on this critical aspect of our neural architecture, we stand to gain a wealth of benefits that extend far beyond the mere joy of connection, touching upon every facet of emotional well-being and cognitive health.

Strengthening Social Bonds for Emotional Mastery

In the dance of human interactions, social bonds serve as the invisible threads that weave us into a coherent tapestry of emotional and intellectual richness. They aren't just niceties of life but are rooted in the core of our biological makeup. The brain, with its astounding complexity, is deeply influenced by these connections. It doesn't merely respond to social interactions; it thrives on them. Strengthening social bonds, therefore, stands as a critical gateway to mastering one's emotions.

Research into the neural underpinnings of social interaction reveals that our brains are hardwired to respond positively to relationships. When we engage in meaningful connections, our brains release a cocktail of neurochemicals—oxytocin, dopamine, and serotonin, among others—that pave the way for emotional well-being. These chemicals act as natural mood enhancers, fortifying our ability to manage stress, enhance happiness, and foster empathy. The art of leveraging these interactions lies in understanding the science of how they affect our brains.

Let's consider oxytocin, often dubbed the "love hormone." This neurochemical plays a pivotal role in social bonding, promoting trust and reducing stress. Engaging in acts of kindness or simply spending time with loved ones can stimulate the release of oxytocin, thereby strengthening our social bonds and facilitating emotional mastery. By nurturing relationships, we tap into a reservoir of neurological resources that enhance our emotional intelligence and resilience.

Interaction doesn't always have to be physical to stimulate these benefits. With advancements in technology, fostering connections through virtual means can also trigger similar neurochemical responses. However, understanding the balance between digital and face-to-face interactions becomes crucial. While virtual connections can be convenient, they must not replace the depth of in-person human interaction that significantly impacts our brain's structure and function.

Creating an environment that prioritizes face-to-face interactions can dramatically enhance emotional mastery. This means investing time in family gatherings, community involvement, or collaborative work settings. Such interactions act as a fertile ground where emotional intelligence can flourish. They provide opportunities to practice empathy, navigate complex emotions, and refine social skills, all crucial for emotional health.

Empathy, as explored in earlier sections, is a cornerstone of social bonding. It involves the ability to understand and share the feelings of another, a process deeply rooted in the brain's mirror neuron system. These neurons are activated not only when we perform an action but also when we observe someone else performing it, making empathy a profound mediator of emotional intelligence. Strengthening these neural pathways through active listening and genuine concern enhances our connections and emotional mastery.

Social connections also contribute to the brain's structural integrity. Studies suggest that individuals with robust social networks tend to have more gray matter in regions associated with emotional regulation and cognitive function. It's a symbiotic relationship where social bonds not only strengthen our mental faculties but are themselves strengthened through cognitive and emotional growth.

Embracing vulnerability is another important facet of deepening social bonds. While it may seem counterintuitive, allowing oneself to be open and vulnerable can significantly strengthen connections and lead to emotional mastery. Sharing fears, dreams, and aspirations with trusted individuals fosters intimacy and emotional understanding, encouraging psychological growth.

Cultivating this vulnerability requires a conducive environment where judgment is replaced with acceptance. It necessitates building relationships with those who value trust and honesty, providing a safe space for authentic exchanges. Such environments not only soothe the psyche but also promote the release of stress-reducing neurochemicals, fueling the brain's capacity for emotional regulation.

For professionals and lifelong learners, understanding these dynamics is essential for personal and professional growth. In the workplace, strengthening social bonds can lead to improved collaboration, enhanced creativity, and a more positive organizational culture. When individuals feel connected and valued, they are more likely to engage deeply, think innovatively, and contribute meaningfully to collective goals.

Emotionally intelligent leaders recognize the importance of these connections. They strive not just to connect at a superficial level but aim to build authentic and trust-rich environments. Such leaders are adept at fostering a culture of inclusivity and empathy, recognizing that the strength of their organization lies in the richness of its social fabric.

Ultimately, strengthening social bonds is about creating a balance. It's about understanding the neurological impacts of our interactions and actively seeking ways to enrich them. It's a call to invest in relationships, recognizing them as a cornerstone of emotional mastery and overall well-being. By harnessing the power of our connections, we unlock pathways to emotional intelligence that are as transformative professionally as they are personally.

Chapter 11:
Mindfulness and the Brain

Mindfulness, an ancient practice rooted in modern science, offers profound effects on the brain's functionality, enhancing emotional balance and mental clarity. Engaging in mindfulness regularly alters the brain's structure; neuroplasticity comes into play, fortifying pathways associated with attention, compassion, and self-regulation. By integrating mindfulness, professionals can experience reduced stress, increased focus, and heightened emotional intelligence, leading to improved interpersonal relationships and optimal performance. As we cultivate these mental habits, the prefrontal cortex strengthens, allowing better management of emotions and a more insightful perspective on challenges. The beauty of mindfulness lies in its accessibility; simple practices breathe life into our cognitive frameworks, nurturing growth and resilience. These transformations, both subtle and profound, highlight the brain's astonishing adaptability, encouraging us to weave mindfulness techniques into daily routines for sustained emotional well-being.

Effects of Mindfulness on Brain Function

Mindfulness is more than just a buzzword; it represents a profound shift in how we understand mental processes.

Recent neuroscience research has given us some profound insights into the effects of mindfulness on brain function, revealing it's not just a tool for stress reduction but a transformative practice with tangible neurological benefits. The brain, with its remarkable plasticity, responds to mindfulness by reshaping itself, which, in turn, affects emotional regulation, attention, and overall psychological resilience.

At the core of mindfulness practices is the principle of being fully present in the moment without judgment. This practice enhances our ability to regulate emotions, as shown by increased activity in areas of the brain associated with emotional processing, such as the prefrontal cortex. The prefrontal cortex is crucial for various higher-order functions, including decision-making and moderating social behavior. When practicing mindfulness, we strengthen this region, which results in enhanced cognitive control and emotional regulation.

Furthermore, mindfulness has been shown to influence the amygdala, the part of the brain involved in processing emotions such as fear and anxiety. Regular mindfulness practice can lead to a reduction in the volume of the amygdala, which correlates with a decreased stress response. The amygdala communicates with various other brain regions to govern our emotional reactions. By shrinking this region, mindfulness helps in moderating the intensity of emotional reactions, aiding in a more balanced emotional state.

Attention, an essential cognitive function, benefits significantly from mindfulness. The practice encourages a focused awareness, which enhances attention control and sustained attention. Studies using functional MRI have shown increased activity in the anterior cingulate cortex, an area of the brain involved in self-regulation and focus, in individuals who engage in regular mindfulness exercises. This heightened activity translates to improved attention span and the ability to stay on task, a valuable skill in both personal and professional environments where distractions are plentiful.

Structural changes in the brain are another fascinating aspect of how mindfulness modifies brain function. Long-term mindfulness practitioners often have increased cortical thickness in brain regions supporting attention and sensory processing. These structural changes are important because they suggest that mindfulness can alter the brain's architecture, contributing to improved cognitive abilities and mental faculties over time. Neuroplasticity allows the brain to adapt and evolve with practice, emphasizing that mindfulness is a skill that gets better with time and consistency.

Mindfulness also impacts the default mode network (DMN), a collection of brain regions active when the mind is at rest and not focused on the outside world. The DMN is often associated with mind-wandering and self-referential thoughts, which can lead to rumination and stress. Mindfulness helps in reducing activity in the DMN, quieting the restless monkey mind, and allowing for greater mental clarity and peace. This reduction provides a pathway to improve internal self-awareness and present-moment orientation, making individuals more attuned to their thoughts and less likely to engage in negative thought loops.

The implications of mindfulness on emotional intelligence are profound. By honing skills such as self-awareness and emotional regulation through mindfulness, individuals become more adept at navigating complex social environments and emotional landscapes. This improvement is linked to increased empathy and better understanding others' emotions, driven by changes in brain regions related to empathy and compassion. As a result, mindfulness practitioners often exhibit enhanced interpersonal skills, critical in both personal relationships and professional contexts.

From a motivational standpoint, the changes brought about by mindfulness practice can inspire individuals to engage more deeply with their personal and professional lives. Understanding that one's brain can change and adapt provides hope and empowerment. These tangible changes in brain function can serve as motivation to continue or begin practicing mindfulness, as they highlight the real, positive outcomes achievable through commitment and practice. Additionally, knowing these benefits are grounded in scientific research can bolster one's resolve to integrate mindfulness into daily life.

To conclude, the effects of mindfulness on brain function demonstrate a powerful intersection between ancient practices and modern science. By understanding these effects, we can appreciate not only how mindfulness shapes our emotional landscape but also how it physically remodels our brains for the better. This insight invites us to consider mindfulness not merely as an occasional exercise but as a foundational tool for lifelong personal growth and professional development.

Mindfulness Practices for Emotional Balance

In the symphony of our minds, where thoughts, emotions, and sensations blend into a complex harmony, mindfulness stands as a conductor that brings order to chaos. This practice, deeply rooted in ancient traditions yet upheld by modern neuroscience, offers a transformative experience for professionals and lifelong learners alike. But what exactly does it mean to weave mindfulness into the fabric of our daily lives, and how can this practice fortify our emotional balance?

The power of mindfulness lies in its simplicity and accessibility. Mindfulness practices involve cultivating a state of active, open attention to the present moment. This can be as simple as focusing on your breath, noticing the sensation of your feet on the ground, or simply being aware of your thoughts without judgment. Such practices seem straightforward, but they engage complex neural pathways known to enhance emotional regulation and self-awareness. Research from cognitive neuroscience has revealed that regular mindfulness practice strengthens connections in the brain's emotion regulation networks, including the prefrontal cortex and the amygdala.

Some practitioners find value in mindful breathing, a foundational practice that centers the mind through the gentle rhythm of inhaling and exhaling. This can be incorporated into the hustle of professional life, with just a few minutes every hour dedicated to focusing on the breath. Such brief interludes can act like a reset button for the brain, clearing the fog of overwhelming emotions and allowing for clearer thinking. Mindful breathing has been shown to reduce cortisol levels, the hormone associated with stress, thereby bringing a calm clarity to decision-making processes.

The beauty of mindfulness is its adaptability. Walking meditation is another technique, less about reaching a destination and more about experiencing the journey. As you walk, notice each step, appreciate the movement, and observe the environment with a keen sense of awareness. This type of meditation can seamlessly integrate into one's commute or even a midday break, offering a graceful escape from the onslaught of stimuli that our brains process daily. Neuroscientists suggest that such practices increase the density of gray matter in brain regions associated with memory, emotion regulation, and learning.

Mindfulness isn't confined to meditation or intentional breathwork. It's a philosophy of presence that can be woven into eating habits, conversations, and even work tasks. Consider mindful eating: instead of hurriedly consuming meals, taking the time to savor each bite can transform a routine task into a visceral, enriching experience. Slowing down and focusing consciously on texture, flavor, and aroma engages sensory pathways while also fostering a sense of gratitude. This practice not only aids digestion but also promotes healthier relationships with food and body awareness.

In a world that prizes multitasking, single-tasking—doing one thing at a time, with full attention—can be a revelation. This approach helps eliminate the stressful scatter-brained feeling that often accompanies juggling multiple tasks. Neuroscientific evidence points to increased productivity and reduced stress when individuals practice this type of mindfulness. The key is to approach each task with intention and purpose, effectively minimizing cognitive load and enhancing task efficiency.

Journaling, when done mindfully, serves as a powerful tool for emotional processing and balance. At the end of each day, setting aside time to reflect on experiences, thoughts, and feelings brings a sense of closure and allows for emotional unpacking. Writing without filters or expectations creates a personal dialogue that can illuminate patterns, enabling better self-understanding. The act of translating feelings into written words activates regions in the brain associated with emotional release and processing, providing insights that sporadic thoughts alone cannot offer.

Incorporating a gratitude practice as part of a mindfulness routine can shift emotional balance towards positivity. Keeping a gratitude journal, where daily reflections focus on moments of joy, support, or learning, can gradually recalibrate the brain's focus on the positive. Studies have found that such practices enhance the brain's response to reward and motivation, fostering emotional resilience and satisfaction.

Moreover, mindfulness practices encourage emotional balance by boosting empathy, a crucial component of emotional intelligence. By training the mind to be present, individuals can better attune to the emotions of others, offering genuine presence and understanding. Activities as simple as mindful listening—truly hearing what another person is saying, rather than planning your response—can vastly improve interpersonal relations. Neural pathways associated with empathy and compassion are activated during such practices, deepening connections and cultivating a supportive social environment.

As mindfulness begins to reshape emotional habits and brain patterns, it fosters a ripple effect across personal and professional domains. Decision-making becomes more thoughtful, stress responses are moderated, and interactions are enriched with authenticity. The mind, like a well-tuned instrument, can perform its symphony with harmony and precision, unencumbered by discordant emotional notes.

Through mindfulness, professionals and lifelong learners can unlock a wellspring of emotional balance, drawing upon both ancient wisdom and modern neuroscience. Emotional turmoil, once a deterrent to clarity and connection, becomes a curiosity to explore rather than a barrier to circumvent. By embracing the mindfulness practices explored here, we step into our potential, cultivating emotional intelligence that fuels both personal fulfillment and professional success.

Chapter 12:
Communication and
Emotional Intelligence

In the dynamic interplay of communication and emotional intelligence, the brain becomes a central player, orchestrating how we connect, perceive, and respond to others. Mastering communication isn't merely about words—it involves decoding the rich tapestry of emotions, body language, and neural cues that govern interpersonal interactions. By leveraging brain-based communication strategies, such as active listening and empathetic engagement, professionals can enhance their capacity to understand and influence. It's about transforming relationships with precision and empathy, guiding dialogues that drive collaboration and innovation. Emotional intelligence acts as the compass, navigating complex social landscapes and bolstering resilience in the face of misunderstandings. As neuroscience unlocks these doors, the opportunity to refine how we convey ideas and foster genuine connections grows, positioning us to thrive both personally and professionally.

Brain-Based Communication Strategies

In the vibrant dance of human interaction, communication acts as the rhythm that lets us connect, understand, and empathize with others. It's not just about exchanging words; it involves tapping into deeper, more intrinsic neurological pathways. Imagine communication as an intricate orchestra, where the brain's neural networks play the symphony. The heart of this concept lies in the brain's ability to weave complex patterns of emotion, cognition, and social understanding, forming the fabric of effective communication.

Our brains are wired to both send and receive a myriad of signals. These signals encompass not just verbal cues, but also nonverbal gestures, emotional expressions, and subtle changes in tone and pitch. All these layers create a nuanced system where effectiveness hinges on the synchronized performance of several brain regions. Central to this process are areas like the prefrontal cortex, which plays a pivotal role in judgment and decision-making, and the mirror neurons that allow us to empathize and connect with others' emotions through subtle mimicry.

The intertwining of emotion and communication is where emotional intelligence takes center stage. Emotion-laden exchanges engage the amygdala, the part of the brain that is often involved in emotional responses, especially those tied to social interactions. It acts somewhat like an emotional radar, scanning for signals that guide our responses. Recognizing these cues and responding with deftness can enhance communication, allowing for deeper connections and the building of trust.

One powerful strategy to enhance communication involves cultivating attentiveness—truly listening and observing without the disruption of preconceived judgments. This attentive state doesn't just involve hearing words but also tuning into the emotional undertones conveyed through facial expressions and body language. Neuroscience reveals that when we focus wholly on the other person, our brains engage in a process called "attunement," allowing us to become more in sync with our conversational partners.

Let's delve deeper into the science. The brain's insula plays a critical role in this process, gathering sensory information and forming a visceral understanding of ourselves and others. The insula helps bridge internal and external experiences, acting as a translator between the two. Enhanced insular activity can lead to a strengthened ability to interpret emotional gestures correctly, making it a key player in improving communicative empathy. As you become more adept, you'll notice this advanced perception helps iron out misunderstandings, fostering clearer dialogue.

Building on this foundation, a practical application arises: addressing how our own state of mind affects communication. Have you ever noticed that your own mood colors your interactions? The brain's limbic system, which governs emotion and memory, significantly influences how we interpret social signals. Learning to regulate our emotions— effectively modulating responses rather than reacting impulsively—can transform the quality of our interactions. Techniques such as mindful breathing or cognitive reframing can help soothe any emotional spikes, allowing for more thoughtful communication exchanges.

In the realm of feedback, brain-based strategies offer a fascinating approach. It's important to create an environment where constructive feedback is welcomed rather than feared. This involves activating our brain's default mode network—a system responsible for self-referential thoughts and internal monologue. Engaging this network fosters introspection and the ability to view feedback not as a personal attack, but as a growth opportunity. By practicing openness and humility, we can cultivate a genuine learning-oriented mindset, enhancing not only our emotional intelligence but also our capacity for more truthful and effective communication.

Further amplifying this process is the prefrontal cortex's role in modulating behavior and responses. As the brain's executive center, it helps oversee complex social interactions, planning, and self-control. Its function can be nurtured through practices such as mindfulness meditation and stress-reduction techniques, which aid in maintaining focus and equanimity during conversations. Developing these skills can lay the groundwork for communication that's not merely transactional but transformational.

What's truly exciting is how the brain can adapt and grow through neuroplasticity. Our communication skills can be honed and refined, with repeated practice reinforcing the neural pathways involved. Imagine communication as a muscle that strengthens with use—this concept is rooted in the brain's ability to form new connections and strengthen existing ones. As you consciously engage in brain-based communication strategies, the synaptic connections involved become more robust, making these new skills more accessible and instinctual over time.

Ultimately, the essence of brain-based communication transforms how we view interaction. It shifts from simply an exchange of ideas to a deeply empathetic engagement with others' realities. The more we connect with others on an authentic level, the more we contribute to a world where communication becomes a means of building bridges rather than erecting barriers.

By embracing these strategies, we can enhance our emotional intelligence to not just understand, but also to influence and inspire others positively. This journey of developing brain-based communication offers both a personal and shared evolution, opening doors to deeper relationships, richer conversations, and an enriched emotional intelligence that empowers us both professionally and personally.

Effective Listening and Understanding Through Neuroscience

Effective listening isn't just about hearing words; it's about comprehending the emotions behind those words. Thanks to advances in neuroscience, we're unraveling how active listening can profoundly influence our interpersonal communication. The secret lies within the brain, particularly in the way neural pathways are dedicated to processing auditory information and emotional signals.

Understanding these complex processes begins with the brain's auditory system. When sound enters our ears, it travels through the auditory cortex. This region is primarily responsible for identifying sounds, differentiating tones, and essentially enabling us to process speech. However, effective listening doesn't stop there. The brain continuously integrates these sounds with contextual and emotional data from other regions, such as the prefrontal cortex and the limbic system, which are vital for emotional processing and decision-making.

To illustrate, consider a conversation between two colleagues discussing a project milestone. The ability to identify not just the words but the underlying frustrations or excitement can transform a simple dialogue into an opportunity for deeper connection. This transformative listening process requires more than just focusing on words; it entails tuning into the speaker's emotion and intention. Neuroscience reveals that this empathetic response activates mirror neurons, which can enhance our understanding by effectively 'mirroring' the feelings of others within ourselves.

The prefrontal cortex plays a significant role in managing these emotional insights. It's where the integration of linguistic, emotional, and cognitive data occurs. This part of the brain helps us make sense of what we hear within the broader context of our thoughts and memories, helping us to understand and respond appropriately. When it's working optimally, we're heightened to cues of empathy and understanding, which are crucial for effective listening.

Moreover, research has shown that oxytocin, often dubbed the 'love hormone,' can enhance our capacity for empathy and understanding during communication. Oxytocin modulates social interactions by fostering a sense of trust and bonding, which are pivotal for effective listening. When we listen with empathy, our brain chemistry shifts toward greater openness and comprehension, making conversations more productive and harmonious.

In practice, incorporating neuroscience into effective listening begins with mindfulness. Being present and attentive in conversation is a foundational aspect. Techniques such as deep breathing and focusing on the present moment can help maintain this mindfulness. These practices engage the brain's attentional networks, enhancing your capacity to focus not only on words but also on non-verbal cues.

Additionally, we must embrace active engagement. This involves not just nodding along but actively participating in dialogue through paraphrasing or summarizing what the other person has said. This process stimulates neural pathways related to memory and comprehension, encouraging a deeper processing of information. It's also important to ask probing questions, not just for clarity but also to demonstrate genuine interest. This validates the speaker's feelings and reinforces neural connections related to empathy and understanding.

Another element to consider is the power of nonverbal communication. Neuroscientific studies highlight the significance of facial expressions, gestures, and tone of voice, which are processed in tandem with verbal content by the brain. Paying attention to these cues can provide invaluable insights into the speaker's emotional state and intentions, further enriching the communication process.

Moreover, developing your emotional vocabulary can stimulate effective listening. By expanding your repertoire of emotional terms, you not only refine your ability to describe your feelings but also enhance your interpretative skills when others express themselves. This nuanced understanding supports the brain's narrative construction process, making emotional interactions more intuitive.

Emotional regulation, another facet of effective listening, ensures that we respond appropriately to the emotional weight of conversations. The ability to manage our own emotions while engaging with others ensures clarity of thought and prevents reactive or defensive responses, which can hinder communication. Engaging the brain's frontal lobes can aid in controlling impulsive reactions and maintaining a balanced perspective.

Lastly, cultivating a growth mindset can significantly aid in developing effective listening skills. By viewing our listening abilities and emotional intelligence as improvable, we tap into neuroplasticity—the brain's ability to rewire itself as we learn and grow. This mindset transforms every conversation into an opportunity for personal and neurological growth.

In summary, leveraging neuroscience for effective listening involves a sophisticated interplay of brain systems and hormones that converge to enrich our understanding and emotional intelligence. By embracing mindfulness, active participation, and emotional regulation, we not only enhance our communication skills but also foster deeper connections with those around us. As professionals and lifelong learners, prioritizing these practices can empower our personal and professional endeavors, ultimately harnessing the full potential of our neural capabilities.

Chapter 13:
Overcoming Emotional Barriers

In our journey to harness neuroscience for personal and professional growth, understanding how to overcome emotional barriers is pivotal. These barriers, often invisible yet palpable, operate as neurological blockages that can stifle our personal and interpersonal development. Identifying these blockages requires a keen understanding of how certain patterns of thought are formed and entrenched within our neural pathways. Once recognized, the key to transcending these limitations lies in employing science-backed techniques—those that encourage rewiring of the brain and adjusting emotional responses. Techniques such as cognitive reframing and mindfulness practices become invaluable tools, allowing us to break free from the shackles of negative emotions and outdated mental constructs. By cultivating a mindset of resilience and adaptability, not only can we surmount these obstacles, but we also foster a deeper awareness that enriches our emotional intelligence. This mastery over our emotional landscape not only propels us forward but also emanates an empowering influence on those around us, enhancing both personal satisfaction and professional relationships.

Identifying Neurological Blockages

In our quest to overcome emotional barriers, identifying neurological blockages stands as a pivotal step. Oftentimes these blockages aren't evident to us, masked by their subtlety and the complexities of the human brain. Understanding the nature of these blockages can unlock pathways to emotional freedom, allowing us to craft healthier neural circuits. But how do we identify these blockages in a brain brimming with billions of neurons constantly firing and rewiring?

Our journey begins with looking into the roots of emotion, residing deep within the brain's architecture. The amygdala, hippocampus, and prefrontal cortex are key players in this intricate dance, each with their own rhythm and role. When any of these areas encounter disruption—be it from trauma, chronic stress, or learned negative patterns—neurological blockages can arise, ultimately hindering our ability to process emotions effectively. Recognizing the signs of these blockages becomes crucial; these might manifest as difficulty managing anger, anxiety spirals, or an overwhelming sense of sadness that seems to linger without resolution.

It's fascinating how the brain's plasticity can be both a boon and a burden. Neuroplasticity is the brain's ability to reorganize itself by forming new neural connections throughout life. However, just as it can be harnessed for growth and healing, maladaptive patterns can also take root. These patterns might lead to emotional blockages, manifesting in unproductive behaviors and thought cycles. By tracing these patterns, we can begin to unravel them. Acknowledging their presence is the first step toward transformation.

To explore further, we delve into how neurological blockages might impact emotional intelligence—our capacity to be aware of, control, and express our emotions, and handle interpersonal relationships judiciously and empathetically. Blockages in emotional processing pathways can diminish our emotional intelligence, leaving us less adaptable to change, less empathetic, and more reactive rather than responsive. Identifying where these interruptions occur can empower us to rectify them, setting the stage for enhanced emotional intelligence.

Patterns of avoidance offer a glaring indication of neurological blockages. When we choose—or rather, when our brain chooses for us—to avoid certain emotions or situations, we effectively reinforce these blockages. Avoidance becomes a cycle; the more we avoid a particular emotional challenge, the stronger the blockage becomes. Whether it's avoiding confrontation out of fear or sidestepping grief through distraction, the neurons strengthen the blockage with each instance of avoidance.

Interestingly, science has begun to illuminate tangible methods to identify these blockages at the neurological level. Techniques such as functional MRI (fMRI) and electroencephalogram (EEG) mapping allow us to visualize areas of heightened or diminished activity in the brain. In therapeutic settings, these tools are used to uncover patterns that might not be evident through introspection alone. By observing how the brain responds to specific emotional stimuli, therapists can help pinpoint where blockages exist and tailor interventions accordingly.

Another approach employs a more introspective lens, focusing on emotional awareness and mindfulness. By cultivating a present-centered awareness of our thoughts and feelings without judgment, we begin to notice our habitual emotional responses and the situations that trigger them.

Mindfulness practices let us observe these blockages from a distance, stripping them of some of their power and beginning the gentle work of dissolution.

As we identify neurological blockages, it's also important to recognize the influence of social and environmental factors. Prolonged exposure to stress or toxic environments can exacerbate emotional blockages, cementing them further into our neural architecture. Acknowledging these external contributors provides a fuller picture of the intricate balance involved in emotional health and highlights the importance of a supportive environment in overcoming these barriers.

Consider the role of sleep and nutrition as well. Poor sleep patterns and nutritional deficiencies can mimic or contribute to neurologically based emotional blockages. Ensuring adequate restorative sleep and a nutrient-rich diet supports optimal brain function, providing the internal resources needed to address and reduce blockages. Making these lifestyle adjustments can facilitate emotional clarity and resilience, paving the way for more effective emotional processing.

In conclusion, identifying neurological blockages requires a multifaceted approach, blending neuroscientific insight with emotional introspection and lifestyle adjustments. By becoming attuned to the signals our brain sends and employing both scientific tools and personal mindfulness strategies, we can dismantle these blockages with patience and persistence. This endeavor not only enhances our emotional intelligence but equips us with the resilience to tackle future emotional challenges head-on. In doing so, we lay the groundwork for more meaningful and fulfilling personal and professional relationships.

Techniques to Break Emotional Barriers

Breaking through emotional barriers can feel much like chipping away at invisible walls. These barriers, often formed over years by deep-rooted fears, past experiences, or even cultural conditioning, prevent us from fully engaging with our emotions or those of others. Luckily, neuroscience offers several techniques that can help dismantle these walls and allow for emotional growth and understanding.

One effective approach starts with self-compassion, a method grounded in neuroscience and supported by evidence indicating how nurturing kindness towards oneself can dampen the negative impact of critical inner voices. Self-compassion activates brain areas responsible for processing emotional pain, enabling us to handle discomfort with more grace and less resistance. This practice shifts the brain's mechanisms, reducing the activity in the amygdala—the brain's alarm system—and increasing the activity in regions linked to empathy and problem-solving.

Engaging in reflective journaling is another potent method. When individuals document their thoughts and feelings, different parts of the brain work synergistically to process emotions constructively. Writing about difficult emotions helps engage the prefrontal cortex, which is essential for regulating emotions and distinguishing between real threats and perceived ones. As you scribble your thoughts, your brain categorizes emotional information, allowing better understanding and diminishing the intensity of those emotions.

An equally transformative exercise involves visualizing positive outcomes. Visualization practices tap into the brain's power to interpret imagined experiences similarly to real ones. By mentally picturing desired circumstances where emotional barriers are overcome, synapses in the brain reinforce pathways aligned with confidence and positivity. This continued exercise contributes to neuroplastic changes crucial for lasting emotional resilience.

Mindfulness meditation stands unrivaled among techniques to break emotional barriers. By focusing awareness on the present moment without judgment, mindfulness enhances emotional regulation. It promotes changes in the brain associated with increased gray matter density in areas related to emotion regulation and self-related processing. Regular practitioners often report greater emotional balance and an enhanced capacity to confront emotionally challenging situations.

Cognitive restructuring also offers a science-backed method to tackle deeply embedded emotional obstacles. This technique involves identifying and challenging distorted thought patterns and irrational beliefs. Rooted in cognitive behavioral therapy, cognitive restructuring helps modify the neural circuitry involved with threat detection and response. By consciously redirecting thoughts towards more constructive pathways, you rewire the brain, making it less reactive to triggers or previously distressing emotions.

Sometimes, tapping into social connections serves as a direct line to breaking emotional barriers. The social brain, a network including regions like the temporal parietal junction and prefrontal cortex, lights up in meaningful interactions, encouraging emotional openness. The oxytocin released during positive social engagement further aids in reducing stress and opening emotional pathways, reinforcing bonds and creating a safe space to express and process emotions.

Physical movement also deserves mention when discussing emotional barriers. Exercise floods the brain with neurochemicals that encourage learning and emotion regulation. Moreover, activities like yoga or Tai Chi incorporate both movement and mindfulness, offering a dual approach to dismantle emotional blockages. As the body moves, the brain's neuroplastic abilities heighten, enabling emotional pathways to adapt and transform.

For lasting change, it's essential to turn these practices into habits. Studies in neuroplasticity affirm that repetition solidifies new neural pathways. Set small, manageable goals for integrating these techniques into your daily routine; consistency over grand gestures is often more effective, as the brain thrives on routine and familiarity.

Facing and breaking emotional barriers isn't just about confronting negative emotions. It involves embracing the full spectrum of human emotion. Techniques that rewire the brain's response to these emotions allow for a richer, more nuanced experience of life itself. As neuroscience continues to evolve, so too will our understanding of how we can leverage these insights for personal and professional growth.

It's about time to start peeling back those layers, examining what's underneath, and choosing methods that align with your unique journey towards emotional mastery. Integrate these scientifically backed techniques and watch the seemingly unyielding barriers give way to a freer, more emotionally connected existence.

Chapter 14:
The Power of Positive Thinking

The human brain, in its intricate design, holds the key to unlocking a more optimistic outlook on life, harnessing the power of positive thinking. By deliberately shifting our focus towards positive emotions, we ignite a cascade of beneficial neurochemical reactions, with serotonin and dopamine leading the charge. This isn't just wishful thinking; it's a scientifically-backed method for enhancing emotional resilience. Cultivating optimism through neuroscience involves consciously rewiring our neural pathways—a process akin to training a muscle to be stronger and more flexible. We're not merely aiming for fleeting happiness, but a deeper resilience that bolsters our capacity to navigate challenges with grace. Through consistent practice, positivity reshapes the brain's structure, making it an ally in both personal growth and professional excellence. As we delve deeper into this chapter, remember that becoming more positive is not about ignoring life's difficulties, but about empowering ourselves to face them head-on with a fortified, resilient mind. Embracing this mindset is a cornerstone for transforming the emotional landscape and paving the way for a more fulfilling, emotionally intelligent life.

Brain Chemistry in Positive Emotion

In the realm of positive thinking, brain chemistry plays a fundamental role in shaping the emotions and experiences that define our lives. At the heart of it all lies an intricate dance of neurotransmitters, those chemical messengers that facilitate communication between brain cells. Dopamine, serotonin, and endorphins are some of the key players, each contributing uniquely to our emotional well-being.

Let's start with dopamine, often called the "feel-good" neurotransmitter. When we anticipate or achieve something rewarding, dopamine levels surge, fueling our motivation and pleasure. This anticipation-reward cycle encourages us to pursue goals, fostering optimism and a positive outlook on life. In essence, dopamine is like a bridge that connects intention with action, providing the drive necessary to realize our aspirations.

Serotonin, another crucial neurotransmitter, is linked to mood regulation and overall well-being. Think of it as a natural mood stabilizer that helps us navigate the ups and downs of daily life. When serotonin levels are balanced, we experience feelings of contentment and calm. Imbalances, however, can lead to mood disorders such as depression, highlighting the importance of serotonin in maintaining a positive emotional state. This neurotransmitter is also intricately connected to sleep, appetite, and digestion, demonstrating its widespread influence over both our mental and physical health.

Endorphins, often associated with the "runner's high," act as natural painkillers and stress reducers. When faced with stress or discomfort, the brain releases these neurochemicals to buffer discomfort, promoting feelings of euphoria and relaxation. This response is especially beneficial in building resilience against stressors, as it allows us to cope more effectively with challenges while maintaining a positive emotional outlook.

The interplay between these neurotransmitters isn't just a matter of binary reactions. It's a dynamic symphony where chemical balances affect how we perceive and react to the world. Consider oxytocin, the so-called "love hormone," which enhances feelings of trust and social bonding. When we connect positively with others, oxytocin reinforces those interactions, encouraging the formation of strong, supportive relationships that elevate emotional well-being.

But how can we harness this knowledge of brain chemistry for personal and professional growth? One approach lies in deliberately cultivating experiences that elevate positive neurotransmitter activity. For instance, setting and achieving small, meaningful goals can trigger dopamine releases, imbuing us with a sense of purpose and accomplishment. Regular physical activity can stimulate endorphins, reducing stress and boosting mood, while mindfulness practices can increase serotonin levels, promoting a serene and balanced outlook.

Yet, the benefits of understanding and leveraging brain chemistry extend beyond individual emotion regulation. In professional settings, fostering environments that promote positivity can enhance collective creativity, efficiency, and problem-solving capacities. Teams that experience higher levels of serotonin, thanks to supportive and appreciative cultures, often exhibit greater cohesion and innovation.

This aspect of positive psychology isn't just about avoiding negative emotions, but rather about amplifying the positive ones. The biochemical pathways that govern our emotions have a profound impact on our mental well-being, directly influencing how we think, feel, and interact with those around us. Engaging in activities that increase neurotransmitter levels, such as social interactions, exercise, or spending time in nature, can lead to enduring enhancements in emotional well-being.

While the biochemical basis of emotion is complex, the essence of using positive thinking to influence brain chemistry lies in the concept of neuroplasticity. This incredible ability of the brain to reorganize itself, forming new neural connections in response to experiences, allows us to nurture emotional growth and resilience. By training our minds to focus on positive stimuli and responses, we can literally reshape our emotional landscape and forge a path toward lasting happiness.

Moreover, integrating these insights into daily routines doesn't require drastic changes. Simple practices such as expressing gratitude, engaging in creative pursuits, or maintaining a balanced diet can have significant cumulative effects on our brain chemistry and emotional health. These small, consistent actions, when coupled with an understanding of the underlying neuroscience, provide a powerful toolkit for enhancing emotional intelligence and overall life satisfaction.

At the intersection of neuroscience and positive psychology, we find a treasure trove of potential for personal transformation. By aligning our behaviors with the brain's natural pathways, we can unlock new dimensions of creativity, resilience, and interpersonal connection. Thus, the science of brain chemistry and positive emotion not only empowers us to improve our well-being but also inspires us to help others on their own journeys of emotional mastery.

In the grander scheme of things, the brain's chemistry is a foundational pillar in the pursuit of joy and fulfillment. As we harness the power of these chemical messengers, we're not just changing our own brains; we're also contributing to a wave of positivity that can influence broader social contexts. By embracing the science behind positive emotions, professionals and lifelong learners alike can forge empowered paths toward more enriched and meaningful existences.

Cultivating Optimism through Neuroscience

In a world brimming with unpredictabilities and challenges, optimism is more than just a desirable trait—it's an essential part of emotional intelligence that drives personal and professional growth. Neuroscience offers fascinating insights into how optimism is not merely a byproduct of a sunny disposition, but rather a result of specific neural processes that can be nurtured and strengthened. By delving into the intricacies of our brain's structure and chemistry, we can understand how optimism functions on a biological level, and more importantly, how we can cultivate it within ourselves.

At the heart of this endeavor is the brain's ability to change and adapt, a phenomenon known as neuroplasticity. This concept underscores the brain's dynamic nature, highlighting that our thoughts and behaviors can, in fact, rewire our neural circuits. When we consistently engage in positive thinking patterns, we strengthen the neural pathways that support optimism. This is akin to building muscle; the more we practice, the stronger our capacity for optimism becomes. The key lies in understanding which parts of the brain are involved and how they interact to produce hopeful outlooks.

The prefrontal cortex, particularly the left hemisphere, plays a significant role in optimistic thinking. This region is associated with goal-setting, decision-making, and the regulation of emotions. By consciously directing our thoughts towards positive outcomes, we activate these areas, which in turn enhances our ability to maintain an optimistic mindset. Additionally, the anterior cingulate cortex helps regulate emotions and manage conflicts between our rational mind and emotional impulses. Practice in focusing on positive outcomes can fortify the connections within this circuit, resulting in a more resilient outlook on life's trials.

Moreover, the release and regulation of neurotransmitters are crucial in fostering optimism. Dopamine, a neurotransmitter linked to pleasure and reward, is instrumental in reinforcing positive experiences. When we anticipate rewarding outcomes or engage in enjoyable activities, dopamine pathways are activated, generating feelings of joy and anticipation. By regularly engaging in activities that increase dopamine levels—such as physical exercise, creative pursuits, or social interactions—we can bolster our propensity for optimistic thinking.

While it may seem simplistic to equate optimism with positive thinking alone, it's essential to recognize its multidimensional nature. Optimism doesn't ignore the reality of life's difficulties; rather, it equips individuals with the courage to confront them, confident in the prospect of favorable results. Neuroscience underscores the fact that optimism requires practice and intentionality. Just as a musician hones their craft through rehearsal, developing an optimistic mindset involves routinely challenging negative thought patterns and reframing them in a more constructive light.

Techniques such as cognitive restructuring and positive visualization can be particularly effective. Cognitive restructuring involves identifying and altering negative thoughts, replacing them with more balanced alternatives. For instance, reframing a perceived failure as a learning opportunity rather than a setback can significantly alter emotional responses. Meanwhile, positive visualization taps into the brain's imaginative power, allowing individuals to mentally simulate positive scenarios, effectively preparing the brain to recognize and pursue opportunities.

What's truly powerful about cultivating optimism is the ripple effect it has on various aspects of life and work. Optimistic individuals are not only better equipped to deal with stress and adversity but also tend to experience improved mental and physical health. This optimistic outlook extends to professional environments as well, where leaders who embody hopeful vision and positive expectancy can inspire and motivate teams, fostering environments ripe for innovation and collaboration.

Furthermore, optimism can enhance social bonds. Neuroscience reveals that the brain's mirror neural system, which plays a role in empathy and understanding, is influenced by our emotional states. Optimism can thus enhance our ability to engage with others empathetically, improving interpersonal relations. This positivity is contagious, promoting a cycle of support and encouragement that benefits entire communities.

In order to nurture optimism effectively, it is crucial to adopt strategies that engage neuroplasticity positively. Mindfulness practices, for instance, have shown considerable promise. By fostering awareness and presence, mindfulness helps diminish the hold of negative biases and habituated responses, providing the brain with space to adopt new perspectives. When paired with gratitude practices, which encourage recognition of positive aspects in daily life, these techniques can profoundly shift neural activity towards patterns associated with optimism.

Ultimately, the cultivation of optimism through neuroscience is both a science and an art. It involves understanding the brain's biological foundations and applying this knowledge through deliberate, consistent practice. By embracing this journey, professionals and lifelong learners alike have the opportunity to not only improve their emotional intelligence but also unlock a pathway to enduring happiness and success.

Chapter 15:
Creativity, Emotion, and the Brain

Creativity and emotion are intricately linked within the mysterious corridors of the brain, where imagination finds its spark and emotional depth lends color and intensity to creative endeavors. At the intersection of neural circuits, our capacity to create is not simply a function of artistic flair but is deeply rooted in our emotional experiences and mental landscapes. Harnessing creativity for emotional insight isn't just about fostering innovation; it's about understanding emotions as they spill into our creative expressions, revealing truths about our inner selves. The brain's remarkable ability to interweave complex emotions with creative processes opens new avenues for personal growth and self-awareness. Encouraging a balance between analytical thought and emotional expression can invigorate creativity, promoting a richer emotional intelligence that lights pathways to both personal and professional evolution. By embracing this synergetic interplay, individuals aren't just cultivating creativity; they're enhancing their ability to navigate the emotional tapestry of their lives, thus progressing toward a more fulfilled existence.

Creative Processes in Emotional Context

The interplay between creativity and emotion is a rich tapestry woven within the complex architecture of the human brain. At the core of our creative processes lie emotions that fuel innovation, ignite imagination, and inspire newfound solutions to both mundane and complex problems. When we dive into the emotional context of creativity, we uncover a landscape where feelings don't just accompany creative endeavors but actively shape and mold them, enhancing the depth of invention and discovery.

Emotions serve as catalysts in our creative processes, each one acting like a different hue on an artist's palette. Consider how the emotion of joy can generate a flood of vibrant, free-flowing ideas, while melancholy might draw out profound, introspective reflections. The emotional framework within which creativity happens is as important as the creative ideas themselves. By understanding this relationship, professionals can not only enhance their emotional intelligence but also leverage it for personal and professional growth.

It's intriguing how the brain processes emotion and creativity in tandem. Emotions originate in key brain areas such as the amygdala and hippocampus, while the prefrontal cortex orchestrates creative thought. These regions do not operate in isolation but engage in a dynamic conversation that inspires creative outputs. People blessed with high emotional intelligence often tap into their emotional awareness, using feelings as a guide to navigate creative problem-solving.

For instance, when faced with a challenging project, recognizing stress and frustration allows for a strategic shift in perspective. This pivot can spark divergent thinking, where seemingly unrelated ideas connect to form innovative solutions. Such emotional awareness isn't an innate talent reserved for a few but a skill that can be nurtured and expanded with practice.

In the workplace, leaders and team members alike can harness their emotional landscape to foster a climate ripe for creative innovation. By encouraging open emotional expression and recognizing diverse emotional contributions, organizations can create environments where creativity thrives. It's no coincidence that the most innovative companies are those with emotionally intelligent cultures that embrace and channel emotion as a source of strength.

Emotionally intelligent individuals often find themselves naturally drawn to creative pursuits, finding that their emotional experiences fuel the creative fires. Yet, everyone possesses the potential to cultivate this connection by deepening their emotional insights. Practices like mindfulness and reflection enhance this emotional awareness, leading to greater creative outputs. They teach us to pause and reflect, creating a mental space where emotions can be acknowledged and transformed into creative energy.

The intersection of emotion and creativity is perhaps most evident in the arts, where emotional expression becomes an integral part of the medium. When artists paint, write, or compose, they're translating emotion into a tangible form that others can experience. This translation is not only therapeutic but also allows the broader audience to connect and empathize, highlighting the universal language of emotion.

Moreover, the creative process acts as an exploration of self, offering insights into the emotional fabric of our being. Emotions provide a window into unmet needs or unresolved issues that can be addressed through creative expression. This cathartic process not only alleviates emotional burdens but also sparks a renewed perspective, opening paths for future innovation.

Innovation and originality thrive in environments where emotions are celebrated rather than suppressed. Think of a brainstorming session where anxiety is openly shared and acknowledged. Such a space fosters creative risk-taking, where fears are reframed not as impediments but as stepping stones to new possibilities. By recognizing the emotional undercurrents that guide our thoughts, we unlock greater creative potential.

In emotionally charged situations, creativity can become a vehicle for problem-solving. When emotions run high, they often obscure rational thinking, but when channelized, they fuel creative alternatives. This is particularly evident during times of crisis, where emotional intensity births groundbreaking innovation. The ability to harness one's emotional energy and redirect it toward constructive outcomes can be empowering, leading to solutions that might otherwise remain undiscovered.

Beyond individual experiences, emotion-driven creativity can unite groups and foster collaboration. By empathizing with others' emotional states, teams can deepen their collective understanding, blending diverse thoughts into a cohesive creative vision. This empathetic approach not only strengthens emotional connections but also enhances the collective creative capabilities of a group.

The ability to manage emotion within the creative process involves a delicate balance. Emotional regulation strategies play a crucial role here, ensuring that emotions enhance rather than hinder creativity. Recognizing when emotions are overpowering and employing techniques to modulate them can transform creative roadblocks into productive pathways.

Ultimately, the relationship between creativity and emotion underlines the intrinsic value of emotional intelligence. As individuals navigate their own emotional landscapes, they unlock a creative reservoir that enriches both their personal and professional lives. Whether you're creating art, developing a new business strategy, or finding ways to connect more deeply with others, it's clear that emotions are not just a backdrop but a vital part of the creative equation.

In this light, we begin to see the expansive potential when emotion is embraced as an ally in the creative process. It's about recognizing that the emotional context isn't just useful but essential for true innovation. As you explore your emotional depths, equipped with insights from neuroscience, you're invited to engage with creativity in ways that have meaningful and lasting impacts.

Enhancing Creativity for Emotional Insight

Creativity isn't just about art or inventing something new; it's a gateway into the deeper reaches of our emotional world. In recent years, the marriage of neuroscience and creativity has birthed compelling insights about their relationship, offering tangible benefits for emotional growth. Harnessing creativity isn't merely about unleashing the artist within. It's about drawing on complex brain processes to better understand and engage with our emotions, providing a valuable pathway toward self-discovery and emotional intelligence.

The brain, with its intricate network of neurons and synapses, has a profound impact on our creative capacities.

Neuroimaging studies reveal a medley of brain regions lighting up during creative tasks, such as the default mode network, responsible for internal thought processes, and the executive control network, which handles focus and decision-making. This interplay is crucial as it facilitates the mind's ability to imagine, analyze, and derive insights from experiences and emotions. By understanding and engaging these networks, we open up new avenues for exploring our feelings and responses to the world around us.

One essential aspect of enhancing creativity for emotional insight involves recognizing the emotional triggers that either stifle or fuel creativity. Emotions such as fear and self-doubt can be hefty barriers, tying down the mind's wandering potential. Conversely, positive emotions like curiosity and excitement can inspire innovation by widening thought patterns and building an encouraging internal environment. The key lies in harnessing these emotions effectively to propel creative thought processes.

To cultivate creativity, it's important to foster an environment that values experimentation and open-mindedness. A mindset that encourages risk-taking and embraces failure as a part of the learning process can accelerate creative output, further enhancing emotional understanding. This dynamic approach empowers individuals to explore emotions more deeply through creative aspects, sparking revelations about personal emotional responses and the broader emotional landscape.

Immersing oneself in creative pursuits such as writing, music, or visual arts can uncover emotions not easily accessible through conventional introspection. The act of creating opens a dialogue between the conscious and subconscious mind, often leading to unexpected insights about personal feelings and motivations. This process turns creativity into a valuable tool for emotional exploration and problem-solving, offering new perspectives on emotional challenges and promoting psychological resilience.

The role of play in creativity and emotion shouldn't be underestimated. Engaging in playful activities stimulates neurological pathways associated with creativity and innovation. This type of engagement boosts mood and lowers stress levels, making it easier for emotional insight to surface. It also nurtures an adaptive mindset capable of coping with change, a crucial component for emotional intelligence in today's complex world.

Collaborative creativity can further enhance emotional insight by connecting with others on a deeper level. Engaging in group creative activities promotes empathy and understanding, as shared creativity invites participants to explore and appreciate diverse perspectives. This interaction broadens our emotional bandwidth, fostering a richer, more empathetic connection with others and enhancing interpersonal emotional intelligence skills.

Mindfulness practices can dovetail beautifully with creativity, serving as a catalyst for emotional insight. By centering oneself and engaging fully in the present moment, mindfulness opens a canvas for creative thought. This combination allows for a clearer vision of one's emotional state and paves the way for innovative solutions to emotional challenges.

Regularly setting aside time for creative endeavors can act as an emotional oasis—a respite that fosters self-reflection and nurtures emotional insight. Scheduled creativity, whether brief or prolonged, creates a habit that encourages ongoing emotional exploration and detachment from daily stresses. This practice not only enhances personal well-being but also continuously sharpens the creative and emotional faculties of the brain.

Moreover, technology offers exciting new frontiers for exploring creativity and emotion. Tools and platforms designed to boost creative collaboration or provide virtual reality environments can expand our creative horizons, offering new perspectives on emotional experiences and approaches to emotional challenges. By intelligently integrating these technologies, we can amplify our creative endeavors, leading to richer emotional insights.

The journey of enhancing creativity for emotional insight is deeply personal and transformative. While the path may vary for each individual, the benefits reaped from traversing it are compelling—ranging from enhanced emotional intelligence to improved mental health and well-being. By nurturing creativity, we invite a deeper understanding of our emotions, providing a foundation for personal growth and a richer, more connected life.

Chapter 16:
Sleep, Emotion, and Brain Function

In the intricate dance between sleep and emotion, the brain orchestrates a nightly symphony that can shape our emotional landscapes. Sleep isn't just restorative; it's a critical time for emotional processing and resilience-building. During REM sleep, the brain actively processes emotional experiences, helping us navigate the complexities of our feelings. Research shows that consistent sleep patterns are linked to enhanced emotional regulation, underlying the importance of sleep for cognitive and emotional recovery. By understanding and optimizing sleep, professionals and lifelong learners can unlock a path to emotional intelligence that is not only sustainable but transformative, fueling both personal growth and professional success. Embracing mindful sleep habits as a cornerstone of brain health empowers individuals to thrive in an ever-demanding world, turning the silent hours of rest into a powerful ally for emotional well-being.

Sleep Patterns and Emotional Health

Our exploration into the intricate web of sleep, emotion, and brain function leads us to a critical junction: the profound impact of sleep patterns on emotional health. The nocturnal realm where we spend a significant portion of our lives holds more sway over our feelings than we may realize. In the dance of neurotransmitters and brain waves, our nightly rest crafts the emotional resilience we exhibit in waking life. Yet, many of us overlook the nuanced art of sleep, underestimating its influence on our minds and emotions.

Understanding sleep's role isn't merely about how much we sleep, but rather, how we sleep. The architecture of sleep is composed of various stages, each with its unique physiological attributes. These stages facilitate restorative processes, where memories are consolidated and emotional landscapes are molded. It's in the interplay of deep sleep and REM (rapid eye movement) sleep that significant emotional processing occurs. REM sleep, in particular, is crucial for processing emotions, as it allows the brain to sort and organize the emotional experiences of the day. Disruptions in these cycles can lead to emotional dysregulation, leaving us feeling irritable and unfocused.

Recent research underscores the bidirectional relationship between sleep quality and emotional health. Poor sleep can lead to heightened emotional reactivity and diminished capacity to regulate emotions effectively. Conversely, a robust emotional state can contribute to more restful sleep by reducing stress and anxiety levels. This cyclical relationship highlights the importance of cultivating healthy sleep habits to maintain emotional equilibrium.

The implications of disrupted sleep extend beyond short-term emotional instability. Chronic sleep deprivation can contribute to longer-term mental health issues such as depression and anxiety. It's a kind of stealthy sabotage, creeping in with fatigue and irritability, quietly eroding emotional balance and resilience. As professionals and lifelong learners committed to personal growth, recognizing and addressing poor sleep patterns is essential for fostering emotional intelligence.

One major factor that influences sleep patterns is stress. The bustling demands of modern life often result in chronic stress, which in turn disrupts natural sleep cycles. Stress activates the body's fight-or-flight response, releasing cortisol, a hormone that can make it difficult to fall asleep. Implementing stress management techniques, such as mindful breathing or guided meditation before bed, can help mitigate stress-induced sleep disturbances, allowing the mind to relax into a state conducive to restorative sleep.

Furthermore, our environment plays a pivotal role in shaping sleep quality. Consider how blue light emitted from screens can trick the brain into perceiving it as daylight, inhibiting the production of melatonin—a hormone necessary for sleep. Establishing a sleep-friendly environment by reducing electronic screen time before bed, maintaining a cool and dark room, and adhering to a regular sleeping schedule can significantly improve sleep quality and, by extension, emotional health.

The concept of sleep hygiene is an essential part of the narrative. It refers to the practices and habits necessary to achieve good sleep quality and full alertness during the day. Developing a bedtime routine that promotes relaxation and consistency can serve as a powerful tool for emotional regulation. This might include activities such as reading, taking a warm bath, or practicing gentle yoga stretches. Regularity in sleep schedules anchors the body's natural clock, helping to stabilize mood and enhance daily function.

Moreover, nutrition can subtly influence sleep patterns as well. Consuming heavy, rich meals too close to bedtime might disturb sleep, while certain foods such as rich in tryptophan, magnesium, and omega-3 fatty acids support the production of serotonin and melatonin, which are key players in sleep and emotional regulation. Building a diet that supports sleep is yet another layer in the multifaceted approach to emotional health.

The interplay between sleep and emotion can also be considered from the perspective of neurobiology. Neurotransmitters such as serotonin and gamma-aminobutyric acid (GABA) are deeply involved in both sleep and emotion. An imbalance in these neurotransmitters not only affects our sleep but also our emotional well-being. Hence, ensuring their levels are optimized through lifestyle choices like diet, exercise, and stress management is indispensable in maintaining both sound sleep and emotional health.

Lastly, it's important to consider the societal perceptions around sleep. In many professional settings, long hours and intense work schedules are often valorized, leading individuals to sacrifice sleep in the pursuit of success. However, this culture of 'burning the midnight oil' can be counterproductive, leading to decreased performance and compromised emotional intelligence. Recognizing sleep as an ally in emotional mastery and success, rather than an obstacle, can reshape how we approach our work and life balance.

As we weave through the tapestry of sleep, emotion, and brain function, it becomes abundantly clear that sleep is not merely a period of rest but a crucial process for emotional resilience and intelligence. By understanding and optimizing our sleep patterns, we can unlock our full emotional potential. It's a call to action, urging us to honor the night as a space for healing and growth in our journey toward emotional mastery.

Improving Sleep for Emotional Recovery

Sleep is not merely a respite from the day's activities; it's a profound state that intricately weaves the fabric of our emotional and mental well-being. Restorative sleep is fundamental for emotional recovery, playing a crucial role in how we process daily stressors, manage emotional memories, and cultivate a resilient mindset. The relationship between sleep and emotional health is a bidirectional highway where each influences the other in a delicate dance of neurochemical symphony.

As we delve into strategies to enhance sleep quality for emotional recovery, we begin by understanding the cycles of sleep. These cycles, comprising rapid eye movement (REM) and non-REM stages, are essential for emotional integration. During REM sleep, the brain is particularly active, processing emotional experiences and consolidating emotional memories. This stage is critical for maintaining an emotional equilibrium as it effectively organizes feelings experienced during waking hours, sweeping away feelings of anxiety and tension.

For professionals and lifelong learners seeking to boost emotional intelligence, the path to better sleep—and thus emotional recovery—begins with establishing a mindful approach to winding down. The brain craves routines that signal the end of the day, calmly transitioning from high alertness to a state of rest. Creating an environment conducive to sleep involves minimizing exposure to artificial light and turning down screens well before bedtime, allowing melatonin production, the hormone pivotal for sleep regulation, to thrive.

The power of consistency should not be underestimated. Cognitive coherence arises from a structured sleep schedule where going to bed and waking up at the same time strengthens our circadian rhythm, leading to more consistent and deeper sleep cycles. As the brain aligns with these natural rhythms, it enhances its ability to process and stabilize emotions, alleviating feelings of mood swings and fostering emotional resilience.

Incorporating relaxation techniques, such as deep breathing exercises or progressive muscle relaxation, can further aid the transition into restorative sleep. These practices not only prepare the body for rest but also significantly reduce cortisol levels, the stress hormone that often acts as a barrier to falling asleep. Lower cortisol levels during sleep allow the central nervous system to fully engage in repair and recovery, diminishing feelings of distress and improving overall emotional outlook.

Another key element to improving sleep for emotional recovery is managing one's diet, particularly before bedtime. The body's relationship with food extends to sleep quality; heavy, spicy meals or excessive caffeine close to bedtime can disrupt sleep, leading to restless nights and heightened emotional sensitivity. Instead, consuming foods rich in tryptophan, magnesium, and B vitamins can promote sleep by facilitating the production of serotonin and melatonin, crucial hormones for relaxation and mood regulation.

Though it is often overlooked, physical activity consistently emerges as a catalyst for improved sleep. Engaging in moderate exercise, preferably earlier in the day, can enhance the quality and duration of nighttime rest. Exercise promotes a smoother transition between the stages of sleep, allowing deeper, more restorative cycles which are vital for emotional processing and recovery. It also acts as an outlet for stress and anxiety, reducing their impact on sleep and emotional health.

The mind's dialogues can often hinder the journey to sound sleep. Practices like journaling before bed help offload worries and promote mental clarity. Reflecting on positive aspects of the day can gear the mind towards relaxation and gratitude, easing the emotional burden that might otherwise intrude into one's sleep cycle. This nightly ritual provides the brain with a safe space to untangle complex emotions and clear the cognitive clutter accumulated during the day.

On the technological frontier, various tools and applications designed to monitor sleep patterns can provide valuable insights. By understanding personal sleep cycles and disruption points, individuals have the opportunity to tailor strategies that specifically address their needs, ultimately leading to improved emotional resilience. Such tools, however, should be used mindfully so as not to become another source of stress or obsession.

In this ongoing pursuit of better sleep for emotional recovery, remember that perfection isn't the goal; consistency and awareness are your partners on this journey. Every small shift towards better sleep hygiene can lead to significant improvements in emotional health over time. As we honor the body and mind's need for quality sleep, we lay the foundation for a more emotionally intelligent and resilient self, ready to navigate both personal and professional landscapes with grace and strength.

Finally, the journey to improve sleep for emotional recovery doesn't just end with tips and techniques—it's about cultivating an overall lifestyle that honors balance. As professionals and lifelong learners, we must recognize that nurturing the brain's capacity for emotional processing through sleep is an investment in personal and professional growth. Embracing a mindset that values rest as much as productivity will lead to better decision-making, enhanced empathy, and a profound depth in emotional intelligence. Sleep, when prioritized and cherished, becomes a powerful ally in harnessing the full potential of the human mind and heart.

Chapter 17:
Nutrition's Impact on
Emotional Well-being

In the tapestry of emotional well-being, nutrition serves as a vibrant thread, weaving through our neural landscapes to foster resilience, clarity, and stability. What we consume directly influences neurotransmitter function and neural circuitry, acting as a critical catalyst for emotional equilibrium. Imagine a diet rich in omega-3 fatty acids, B vitamins, and antioxidants; these nutrients support synaptic plasticity and enhance mood-regulating systems. By choosing foods that boost serotonin production and reduce inflammation, we pave the way for improved mood regulation and cognitive performance. The journey towards optimal emotional health is challenging, but conscious nutritional choices empower us to amplify our brain's potential, unlocking paths to greater emotional insight and mastery. Consuming nutrient-dense, whole foods not only nourishes the body but also sharpens the mind, creating a synergy that propels us toward emotional fortitude and personal growth. Let nutrition be the deliberate architecture of your emotional resilience, building the foundation for thriving minds and fulfilling lives.

Brain-Boosting Nutrients for Emotional

Health

Within the intricate web of our emotional well-being lies a profound connection to the nutrients we consume. Our brains, robust yet remarkably sensitive, rely heavily on a steady supply of vitamins, minerals, and other compounds to function optimally. It's akin to fueling a high-performance vehicle; the quality of the fuel dramatically influences performance and longevity. To understand this relationship is to wield the power to enhance emotional intelligence, boost resilience, and increase overall well-being.

Start with omega-3 fatty acids, essential fats not naturally produced by the body but crucial for brain development and function. Found in fish like salmon, mackerel, and sardines, these fatty acids play a pivotal role in the formation of cell membranes in the brain, influencing synaptic plasticity—the ability of synapses to strengthen or weaken over time, essential for learning and memory. By integrating omega-3s into your diet, you're effectively promoting emotional balance and elasticity of mood, reducing the risk of depression and anxiety.

Alongside these beneficial fats, antioxidants such as vitamin E, found prominently in nuts and seeds, protect our cells from oxidative stress. Oxidative stress is akin to the slow rusting of machinery—it damages cells, proteins, and DNA. The brain, being particularly susceptible, stands to benefit significantly from the inclusion of antioxidants. Through their protective action, antioxidants help maintain the integrity of brain cells, support neural function, and prevent cognitive decline.

Similarly, B vitamins—often heralded as the building blocks for a healthy mind—are indispensable for emotional stability and brain health. Vitamin B6 aids neurotransmitter synthesis, literally facilitating communication between nerve cells. B9 (folate) and B12 are crucial for the breakdown of homocysteine, an amino acid linked to increased risk of neurodegenerative disorders when present in high levels. Leafy greens, beans, and fortified grains are rich sources that, when consumed regularly, can enhance mood, sharpen concentration, and heighten energy levels.

Magnesium, often underappreciated, quietly supports over 300 biochemical reactions in the body. Its role in brain function is profound, influencing mood and emotion by regulating neurotransmitter release and acting as a natural muscle relaxant, which can alleviate tension and anxiety. Foods rich in magnesium include whole grains, dark chocolate, and leafy greens, each promising a calming effect on the nervous system—a much-needed counterbalance to the frenetic pace of modern life.

The amino acid tryptophan also deserves mention, particularly for its role in the synthesis of serotonin, the neurotransmitter commonly associated with feelings of happiness and well-being. Tryptophan finds a home in turkey, bananas, and cheese, and by participating in serotonin production, it can mitigate symptoms of depression, anxiety, and insomnia. A seemingly indulgent plate of pasta with a turkey breast could, in reality, be a strategic play in your emotional well-being arsenal.

Amidst the clamor for mainstream superfoods, turmeric stands prominently with compelling claims backed by emerging research. Curcumin, the active compound in turmeric, is known for its anti-inflammatory and antioxidant properties, which help combat inflammation within the brain—a potential precursor to depression and other mood disorders. Regular inclusion of turmeric in meals or as a supplement could offer a substantive mood boost, enhancing cognitive health.

The gut-brain axis, a burgeoning field of interest, reveals the profound interplay between digestive health and emotional well-being. Probiotics, friendly bacteria found in yogurt, kefir, and fermented foods, can influence the production of neurotransmitters, thus affecting mood and cognition. By cultivating a healthy gut microbiome, you're potentially easing anxiety and depressive symptoms, underscoring the wisdom of the adage "you are what you eat."

Focusing on hydration, water, the most basic yet essential nutrient, often gets sidelined in conversations about nutrition and emotional health. Dehydration can impair concentration, increase feelings of irritability, and even contribute to longer-term emotional dysregulation. Ensuring adequate water intake is not just a recommendation for physical health but a cornerstone practice for maintaining emotional equilibrium.

Caffeine, while ubiquitous, is a double-edged sword when it comes to emotional health. In moderation, it's a stimulant that promotes alertness and enhances mood. However, excessive consumption can lead to anxiety and disrupt sleep patterns—fundamentally undermining emotional well-being. A balanced intake, mindful of timing and quantity, is advised to harness benefits while mitigating downsides.

As you can see, adopting a diet rich in these brain-boosting nutrients requires knowledge and intentionality but promises returns in emotional vigor and resilience. It transforms eating into an act of empowerment, placing the individual in the driver's seat of their emotional health journey. While no single nutrient or food guarantees complete emotional harmony, the cumulative effect of dietary choices undoubtedly paves the path toward heightened emotional intelligence.

Ultimately, think of your dietary habits as foundational to a vibrant and stable emotional state. By embracing nutrient-rich foods, you're not just nourishing your body—you're fundamentally enhancing your brain's ability to process emotions. This endeavor, nurtured by a continuous commitment to nutritional excellence, can foster an emotionally intelligent mind, primed to navigate the complexities of personal and professional life with grace and acuity.

Integrate these principles today, and you're setting the stage for a life marked by enriched emotional health. Your brain is not merely a beneficiary of this nutrition-centric approach; it's an active participant, evolving and adapting in symphony with the nourishing inputs it receives. Through food, you can sculpt a brain resilient in its emotional capabilities, echoing the broader goal of this journey—personal and professional transformation deeply rooted in the science of nutrition and brain health.

Creating a Diet for Optimal Brain Function

In the vast symphony of our emotional well-being, nutrition plays a role that often goes unsung. A well-crafted diet doesn't only fuel the body; it nurtures the mind, laying the foundation for optimal brain function. For professionals and lifelong learners, understanding the intricate relationship between diet and brain health can be transformative. By prioritizing foods that bolster cognitive abilities and emotional health, one can harness a diet to promote emotional intelligence and interpersonal skills.

Consider the impact of omega-3 fatty acids, the much-lauded nutrients with the edge of scientific backing. Found generously in fatty fish like salmon, mackerel, and sardines, these lipids are essential for maintaining brain cell membrane fluidity. When your brain cells communicate smoothly, it translates to improved mood and cognitive function. Omega-3s have been shown to reduce anxiety and depression, conditions that can hinder emotional regulation and self-awareness. Incorporating a serving of fish into your weekly meals is a simple yet effective step toward a sharper mind and balanced emotions.

The brain thrives on glucose, its primary source of energy, yet not all sugars are created equal. Refined sugars can lead to erratic energy levels and mood swings, sabotaging interpersonal interactions and decision-making. Instead, complex carbohydrates, such as those in whole grains, offer a steady release of energy. With options like oats, brown rice, and whole wheat, these carbohydrates provide active sustenance that enhances concentration and emotional stability. Your morning oatmeal could be the unsung hero of your day's productivity and emotional resilience.

Among nature's hidden treasures, antioxidants stand as warriors against oxidative stress, which can impair brain function. Berries, particularly blueberries, are rich in flavonoids that cross the blood-brain barrier, enhancing communication between neurons and promoting memory. Studies have indicated that a diet rich in antioxidants correlates with improved mood and cognitive performance. A simple addition of a berry smoothie or a handful of blueberries into your routine can invigorate your brain and uplift your emotional state.

Our brains thrive on a delicate balance of neurochemicals, and some nutrients play pivotal roles in synthesizing these critical components. Consider tryptophan, an amino acid found in turkey, eggs, and cheese. Tryptophan is a precursor to serotonin, a neurotransmitter that governs mood and social behavior. A deficiency in serotonin can lead to depressive symptoms and a decrease in empathy, crucial aspects of emotional intelligence. Including tryptophan sources ensures your brain has the building blocks it needs to foster emotional harmony.

Hydration is another cornerstone of cognitive clarity. Dehydration can lead to fatigue, confusion, and anxiety - all barriers to effective communication and emotional regulation. While water should remain a primary source, beverages like green tea introduce additional benefits. Rich in L-theanine, green tea can enhance alertness and mental clarity without the jitters associated with coffee. Embarking on your day with a cup of green tea may heighten your mental acuity and maintain your emotional equilibrium throughout stressful situations.

For professionals striving to harness their brain potential, magnesium offers an avenue worth exploring. Found in dark leafy greens, nuts, seeds, and bananas, magnesium acts as a buffer against the physical repercussions of stress. It supports neuroplasticity - the brain's ability to adapt and reorganize itself - which is vital for emotional growth and resilience. As you face daily pressures, incorporating magnesium-rich foods can serve as a protective layer against burnout, allowing you to manage emotions with grace and composure.

Incorporating mindfulness into mealtimes enriches the relationship between diet and brain function. Being present while eating encourages better digestion and absorption of nutrients, amplifying their benefits. This practice of mindful eating can reduce stress and enhance food's role in emotional well-being. Taking a moment to savor your meals allows you to connect deeply with the foods that are actively supporting your mental and emotional state.

While dietary choices fundamentally support brain health, balance and moderation remain key. The occasional indulgence does not derail the pursuit of optimal brain function; instead, it emphasizes the importance of a sustainable, enjoyable approach to nutrition. Just as the brain requires balance in neurotransmitters and hormones, so too does maintaining a varied diet ensure comprehensive nutrient intake.

As professionals striving for excellence, understanding nutrition's impact on emotional well-being can offer new pathways to personal and professional growth. Food is not merely a sustenance of the body; it's a catalyst for emotional depth and cognitive prowess. By choosing foods that support brain health, we arm ourselves with the neurons, neurotransmitters, and neurochemicals essential for emotional intelligence. Our diet becomes a partner in our journey toward enhanced self-awareness, empathy, and emotional resilience. Through the symbiosis of nutrition and neuroscience, a diet for optimal brain function propels us toward our fullest potential.

Chapter 18:
Physical Activity and
Emotional Mastery

Harnessing the synergy between physical activity and emotional mastery opens a dynamic pathway to optimal emotional and cognitive well-being, a dance choreographed by the brain's miraculous adaptability. When we engage in exercise, not only does it spark the release of essential neurotransmitters like endorphins, serotonin, and dopamine, but it also stimulates neurogenesis, fostering the growth of new brain cells that bolster resilience and emotional stability. This biological ballet supports an enhanced mood while tempering anxiety and stress, creating a fertile ground for emotional insight and mastery. Implementing simple, enjoyable physical activities into daily routines, whether it be brisk walking, cycling, or dance, empowers individuals to manage emotions more effectively, ultimately leading to greater interpersonal effectiveness and emotional intelligence. As these movements become habit, they transform the landscape of the brain, opening doors to personal and professional growth through the subtle yet powerful mastery of emotion.

How Exercise Affects Brain and Emotion

Physical activity, often hailed as a cornerstone of health, goes beyond its apparent benefits to the body. It's a potent catalyst for profound changes in the brain, intricately affecting our emotions. Regular exercise, whether it's as simple as a brisk walk or as intense as a marathon run, has been shown to enhance cognitive functions and emotional well-being. The secret lies in our brain's adaptability and its incredible relationship with physical movement.

The brain is an extraordinarily adaptable organ. At the heart of this adaptability is neuroplasticity, the brain's remarkable ability to reorganize itself by forming new neural connections throughout life. Exercise serves as a powerful stimulus for neuroplasticity. When you engage in physical activity, you prompt the release of several growth factors that affect the health of brain cells, the growth of new blood vessels in the brain, and, crucially, the abundance and survival of new brain cells, stimulating both neurogenesis and synaptic plasticity.

Moreover, aerobic exercise has been repeatedly shown to elevate levels of brain-derived neurotrophic factor (BDNF), a protein with a critical role in the survival and growth of neurons. Increased BDNF levels have been linked to better mood, enhanced learning, and sharper memory. Regular physical activity, therefore, is akin to a mental fertilizer, nurturing the growth and connectivity of neurons in areas critical for emotional processing and cognitive functioning.

Exercise also fosters emotional mastery by modulating the chemicals in the brain responsible for mood regulation. Endorphins are often highlighted for their mood-boosting properties, and indeed they play a role in the so-called "runner's high," but they are just a part of a more complex neurochemical ballet. Physical activity ramps up the production of serotonin and norepinephrine, neurotransmitters that help alleviate depression and anxiety. Dopamine, another crucial neurotransmitter associated with feelings of pleasure and satisfaction, is also enhanced by regular exercise, which can improve motivation and focus.

The prefrontal cortex, responsible for executive functions such as decision making, emotional regulation, and goal planning, is particularly sensitive to the positive influences of exercise. When you maintain a consistent exercise regimen, you help in fortifying this brain region's ability to manage stress and cultivate emotional resilience. It's akin to strengthening a muscle—you become less reactive, better at handling emotional challenges, and more adept at maintaining a positive outlook in the midst of adversity.

Interestingly, exercise has transformative effects on areas of the brain associated with stress response, like the amygdala. This almond-shaped cluster of nuclei plays a crucial role in how we process emotions and react to threats. Regular physical activity has been found to decrease amygdala size and reactivity—essentially dialing down the sensation of stress and enhancing emotional control, providing a buffer that mitigates anxiety and fear responses.

Social interactions often accompany physical activities, and these interactions can have an additional positive impact on our emotions. Being involved in team sports or group exercise classes can provide a sense of community and belonging, further enriching the emotional benefits of exercise. The oxytocin released in social settings during group exercise enhances well-being, connection, and trust among participants, fostering a positive outlook and reinforcing emotional ties.

Physical activity doesn't just cultivate positive emotions; it also helps in reducing symptoms of negative moods and conditions. Individuals with chronic mood disorders, including depression and anxiety, often find significant relief after adhering to an exercise routine. While exercise is not a stand-alone treatment, it is a valuable complement to therapy and medication, improving symptoms and enhancing the overall quality of life.

Adding regular physical activity into daily life also benefits sleep patterns, which in turn impacts emotional well-being. Exercise has been shown to help regulate circadian rhythms, promote deeper sleep cycles, and reduce insomnia symptoms. Quality sleep is integral to emotional mastery because it's during sleep that the brain processes emotions experienced during the day, consolidating emotional memory and reducing emotional reactivity.

It's essential to approach exercise with a mindset that aligns with personal goals and capacities. The type, intensity, and duration of exercise can be tailored to individual preferences. Some may find solace in yoga or tai chi, activities that combine physical exertion with meditative practices, enhancing both physical and mental health. Others might prefer the adrenaline rush of high-intensity interval training or the tranquility of a long nature walk. The key is consistency and finding joy in the movement, which in turn, maximizes its emotional benefits.

To truly master emotions through physical activity, it is important to set realistic goals and gradually build a routine. Integration is crucial; making small, sustainable changes can have profound effects over time. For many, the journey starts with short walks and gradually builds to more engaging regimes that challenge not just the body, but the mind and spirit.

Connecting exercise to emotional well-being allows us to use movement as more than just a means to an end. It's a journey towards self-improvement that spans body and mind. Embracing exercise as a way to fortify brain health and emotional resilience empowers us to navigate life's challenges with agility and insight.

Exercise is an investment in ourselves, promoting a cycle where physical activity begets mental clarity, emotional stability, and an enriched quality of life. As you commit to integrating more movement into your day, you're harnessing a powerful tool for emotional mastery through one of the simplest yet most effective methods available. Through the lens of neuroscience, we understand that each step, each heartbeat, and each breath taken during exercise reconstructs not just our bodies, but our brains and emotions, leading us to live richer, more fulfilled lives.

Implementing Physical Activity for Emotion Management

Physical activity offers more than just physical benefits; it's a powerful tool for managing emotions effectively. The link between movement and mood is not new, yet its potential is often underestimated. A robust body of research supports the notion that exercise influences emotional health profoundly, encouraging practitioners and learners alike to incorporate it into their lives intentionally. The key lies not in understanding exercise as a mere physical exertion but as a transformative emotional experience that rewires the brain, enhances neuroplasticity, and fosters emotional proficiency.

Engaging in regular physical activity triggers biochemical changes in the brain. When you move, your brain releases a medley of neurochemicals such as endorphins, serotonin, and dopamine. These neurotransmitters play a crucial role in elevating our mood, acting almost like nature's antidepressants. Endorphins, commonly known as the "feel-good" hormones, act as natural painkillers and mood enhancers. Simultaneously, serotonin helps regulate mood and social behavior, and dopamine reinforces feelings of pleasure and reward. Together, they create a powerful cocktail that enhances emotional well-being.

Interestingly, physical activity also encourages growth in the brain regions associated with emotion regulation. Exercise promotes neurogenesis, particularly in the hippocampus, a brain area integral for emotional regulation and memory. As new neurons are forged and connections strengthened, individuals often experience improved emotional processing capabilities, better stress management, and enhanced resilience to emotional challenges. It's as though the brain is continuously being rewired, setting the stage for emotional growth and mastery.

For those seeking emotional intelligence, regular exercise can transform how they interact with their emotions. Physical activity can act as an outlet, allowing pent-up energies and frustrations to be spent in a constructive way. Activities like running or cycling can serve as a form of moving meditation, creating a balance between body and mind. This dynamic synergy fosters mindfulness and self-awareness, improving one's ability to recognize and understand their emotional states without judgment. As mindfulness practices often rely on stillness, integrating movement adds a novel dimension, helping those who might struggle with traditional static forms of meditation.

Diverse forms of physical activity offer varying emotional benefits, making it important to tailor one's regimen according to personal preferences and emotional needs. Aerobic exercises like swimming or dancing might inject a sense of joy and liberation, while strength training can instill a profound sense of empowerment and confidence. Even activities that emphasize precision and balance, such as yoga and tai chi, nurture a sense of tranquility and centeredness, equipping individuals with tools to handle emotional fluctuations with grace. The goal isn't perfection or competitiveness, but the intrinsic joy of movement.

Implementing physical activity for emotion management requires intention and consistency. Start by setting realistic and sustainable goals. Whether it's a brisk walk in the park or a vigorous session at the gym, the idea is to cultivate a routine that fits seamlessly into your lifestyle. Motivation may wax and wane, but envisioning the emotional benefits could provide the extra push to maintain that routine. Consider joining community classes, or finding a workout buddy, these social components can amplify the emotional gains through connections and shared experiences.

Moreover, integrating physical activity with other emotional management techniques amplifies the overall benefits. Pairing exercise with creative visualization or positive affirmation becomes a potent regimen for emotional mastery. As you engage in physical activity, visualize emotional barriers dissolving or repeat affirmations that reinforce your emotional goals. The combination of movement, visualization, and positivity catalyzes emotional healing and creates an upward spiral of emotional mastery.

Tracking progress in emotional management through physical activity is another layer of this multi-faceted approach. Journaling about emotional states before and after exercise sessions can provide insights into patterns and progress. Recognizing improvements in mood, stress levels, and overall emotional health can be motivating and affirming. At times, journaling might reveal unexpected emotional triggers or breakthroughs, offering valuable learning experiences in the journey toward emotional intelligence.

Finally, it's important to acknowledge that emotional management is a continuous journey rather than a destination. Integrating physical activity isn't about achieving a permanent emotional state but honing the adaptability to navigate life's challenges. Some days will be harder than others, yet it's in those challenging moments where physical activity often shines as a steadfast ally. Remember, each session reinforces neural pathways, empowers emotional regulation, and charts the path toward greater emotional intelligence.

The transformative potential of physical activity for emotion management is rooted not just in movement itself, but in what movement represents—a commitment to self-enrichment and a step towards emotional mastery. As individuals and professionals embark on this path, they cultivate not only a healthier body but also a resilient mind, capable of navigating the complexities of emotion with insight and grace. Thus, the act of propelling oneself forward physically translates to a deeper, metaphorical propulsion towards emotional growth and understanding.

Chapter 19:
Overcoming Emotional
Dependencies

I n the intricate dance of human emotion, dependencies often act like chains, tethering us to patterns that undermine our growth. Understanding the neurological underpinnings of these emotional habits is crucial for liberation. The brain, with its plasticity, offers pathways to break these bindings and forge new, healthier connections. By identifying triggers and reprogramming neural circuits—practices rooted in science-backed techniques—we can transcend old dependencies. It's about repatterning emotional responses through deliberate and consistent effort, leveraging the brain's ability to adapt. As we cultivate self-awareness and resilience, we nurture the seeds of autonomy and fulfillment. This journey demands commitment and a willingness to explore vulnerabilities, yet it's through this challenging process that empowerment emerges. Harnessing the power of neuroscience, we're capable of transforming emotional chains into ladders of personal and professional evolution.

Neurological Aspects of Habits and Dependencies

Our brains are remarkable organs, so adaptable and wired to form patterns that guide us through life with efficiency. When it comes to habits and dependencies, these patterns become particularly pronounced, intertwined deeply with our neural pathways. Understanding the neurological underpinnings of these patterns sheds light on how they influence our emotional dependencies and provides a scientific basis for overcoming them.

From a neurological perspective, habits form because our brains seek shortcuts to conserve energy. The basal ganglia, a group of nuclei in the brain associated with controlling voluntary motor movements and routine behaviors, play a pivotal role in habit formation. They take complex actions and simplify them into automatic responses. This efficiency allows us to perform tasks without expending much cognitive effort.

Dependencies, whether they're behavioral or related to substances, often evolve from habits. The neural circuitry related to reward and pleasure, primarily involving dopamine, significantly contributes to these dependencies. When we experience something pleasurable, the brain releases dopamine, creating a feeling of reward. Over time, our neural pathways become conditioned to seek out behaviors or substances that trigger this rewarding feeling, reinforcing the dependency loop.

While the basal ganglia streamline habits, the prefrontal cortex—a region involved in decision-making and impulse control—can counteract these automatic behaviors. However, dependencies can weaken its influence. For someone engulfed in an emotional dependency, whether it's attachment to a relationship or reliance on external validation, the urge to maintain the status quo often overshadows rational decision-making. This is due to the hijacking of the brain's reward system, making the prefrontal cortex struggle to assert itself.

Neuroplasticity, the brain's ability to reorganize itself by forming new neural connections, offers hope for those aiming to overcome emotional dependencies. It suggests that by introducing new, healthier habits and reinforcing them, we can effectively rewire the brain. This concept dovetails with the idea that our brains are never static; they're dynamic and continue to evolve based on our actions and experiences.

Research into addiction offers valuable insights into the mechanics of dependencies. Both substance addictions and behavioral dependencies show that the more you engage in a specific behavior, the more entrenched the neural pathways become. Just as water carves a deeper groove in rock over time, repeated behaviors solidify these pathways, making change difficult but not impossible.

Studies on mindfulness and meditation highlight their potential to alter brain function and structure favorably. Regular mindfulness practice increases activity in the prefrontal cortex, enhancing its regulatory control over habit loops managed by the basal ganglia. Meditation also nurtures conditions for neurogenesis, the birth of new neurons, which supports brain plasticity.

Breaking free from emotional dependencies involves creating a new roadmap for your brain. This process begins by identifying triggers and substituting old habits with positive routines. Consider the implications of working memory—the system that holds information temporarily for processing. Enhancing working memory through cognitive exercises or mindfulness can improve focus and reduce the mind's tendency to default to ingrained dependencies.

Engaging in physical exercise is another tactic supported by neuroscience. Exercise elevates levels of brain-derived neurotrophic factor (BDNF), a protein that promotes the growth of new neurons and synapses. This biochemical boost aids in neuroplastic changes, providing a physical foundation to support the mental shift away from dependencies.

Furthermore, social connections play a crucial role. Mirror neurons in our brains fire not only when we perform an action but also when we observe another person doing the same. This neural mirroring allows us to learn through social interactions and relationships. Cultivating supportive, understanding relationships can challenge the neural foundations of emotional dependency, providing motivation and reinforcement through empathetic understanding and encouragement.

It's also essential to consider the role of self-awareness in regaining control. Being aware of one's own emotional states, understanding triggers, and recognizing patterns of behavior enable the individual to step back and reflect rather than react impulsively. Enhancing self-awareness can be facilitated by journaling, therapy, and mindfulness practices. As self-awareness increases, so does the capacity to make conscious choices that align with desired emotional outcomes.

Ultimately, becoming more emotionally self-sufficient involves a deep dive into our brain's reward systems, recognizing the influence of habit loops, and systematically altering our environment, thoughts, and actions to reshape them. By leveraging our understanding of neuroscience, we empower ourselves to adjust the neural mechanics governing our behaviors, offering a clear pathway to overcome emotional dependencies.

Strategies for Overcoming Dependencies Using Brain Science

Dependencies, whether emotional, behavioral, or chemical, often find their roots entrenched in complex neural networks and brain pathways. Understanding the science behind these can help us craft effective strategies to overcome them. The brain's capacity for change—its neuroplasticity—offers a beacon of hope, suggesting that with targeted interventions, even deeply ingrained dependencies can be reshaped or dissolved.

At the core of overcoming emotional dependencies is the concept of neural remodeling. This concept leans on the principle that our brains are dynamic and can form new connections in response to learning, experience, or intervention. When we repeatedly engage in a dependency-driven behavior, it strengthens specific neural pathways, making it challenging to alter our habits. However, brain science suggests that with perseverance and the right strategies, these pathways can be weakened and new, healthier ones can be built.

One effective way to initiate this transformation is through mindfulness-based strategies. Mindfulness practices enhance self-awareness, allowing individuals to recognize triggers that precipitate dependency behaviors. By focusing one's attention on the present moment and acknowledging thoughts and emotions without judgment, mindfulness can gradually weaken the automated responses rooted in dependency. Studies have shown that regular mindfulness practice can actually decrease gray matter density in areas of the brain associated with stress and dependency, aiding in emotional regulation and reducing the hold of dependencies.

Another crucial element in overcoming these dependencies is cognitive restructuring. This technique involves identifying and challenging the irrational beliefs that fuel dependencies. Our brain is adept at creating stories around why we need something or why we can't let go. By actively engaging in cognitive restructuring, we can develop healthier, more rational thought patterns that don't support dependency-driven behaviors. This switch not only aids in emotional healing but also encourages the brain to lay down different neural pathways that support autonomy and resilience.

The strategic use of behavioral conditioning is also highly effective. Leveraging the brain's reward system, we can reprogram our responses to certain stimuli. By associating positive experiences with new, healthier behaviors through techniques such as positive reinforcement, we teach the brain to seek out and repeat these beneficial patterns. Neurochemicals like dopamine play a significant role here, reinforcing behaviors that yield rewards—emotional or physical—reinforcing the shift away from dependencies.

Social connections and supportive environments serve as vital tools in overcoming dependencies. Oxytocin, often called the 'love hormone,' strengthens bonds and provides emotional support, which can act as a buffer against dependency. Being in the company of empathetic individuals can help reduce feelings of loneliness and isolation that often accompany dependencies, and provide the encouragement needed to pursue change. By fostering an environment that promotes safety and understanding, the brain can more effectively engage in the process of healing and overcoming emotional dependencies.

Another powerful strategy involves harnessing the brain's frontal lobe. This part of the brain is crucial for decision-making, problem-solving, and impulse control—all necessary functions when overcoming dependencies. Strengthening this area through practices such as meditation, puzzles, or any activities that challenge the mind can improve our ability to resist dependency cravings and make conscious, healthy decisions. Developing a robust capacity for such mental gymnastics empowers individuals to navigate the complex emotions and impulses that often accompany dependencies.

Physical activity can't be overlooked as a tool for dependency management. Exercise has been proven to stimulate the production of neurotrophic factors, which support the growth of neurons and the synapses that connect them. This neurobiological benefit aids in repairing the brain's circuitry affected by dependency. Moreover, the endorphins released during physical activity can lift mood, reduce stress, and help satisfy emotional needs that might otherwise fuel dependency-driven behaviors.

Education remains a cornerstone in overcoming emotional dependencies. By demystifying the workings of the brain through neuroscience literacy, individuals can better understand the science behind their dependencies. Knowledge empowers, enabling people to identify points of intervention and to employ customized strategies that align with their personal journey toward overcoming dependencies. An informed approach increases self-compassion, reduces stigma, and creates a mindset receptive to change.

However, the most fundamental element in all these strategies is time. The process of overcoming dependencies isn't instantaneous; it thrives on consistent practice and gradual progression. The brain's plasticity means that though change is possible, it is often incremental and may require steadfast commitment and repeated efforts. Setting realistic goals and celebrating small victories nurture motivation while cultivating patience and perseverance.

Lastly, professional guidance can play a significant role in this journey. Behavioral therapists or counselors trained in neural techniques can provide the tailored support and structured interventions necessary to deconstruct dependencies at a neural level. Working with a professional allows for accountability and offers the opportunity to use cutting-edge neural sciences tools tailored to individual needs.

In summation, overcoming emotional dependencies using brain science is a journey of re-wiring, retraining, and re-facing the challenges that come with deeply entrenched habits. The interplay between understanding our brain's capabilities and implementing targeted interventions serves as a potent formula for change. As we look inward and harness the power of our neural networks, the path to independence becomes not just possible, but sustainable and enriching. The transformation forged through these strategies unfolds a future wherein our emotional dependencies no longer dictate our lives, but rather our brains steer us toward healthier, more fulfilling paths.

Chapter 20:
Decision-Making and
Emotional Influence

In the multifaceted world of decision-making, emotions wield an often underappreciated yet profound influence. It's not just about making choices that appear rational; the brain's emotional circuits play an intricate dance with cognitive processes, guiding our decisions in ways we might not fully comprehend. When you're faced with a decision, whether personal or professional, your brain taps into a rich tapestry of emotional memories and subconscious biases that subtly sway your choices. By harnessing these emotional undercurrents with greater awareness, professionals can transform what seems like an art into a science, leveraging emotional insights to enhance the quality and impact of their decisions. Cultivating emotional intelligence allows us to see with clarity, not just through logic, but with empathy and intuition, bridging the gap between raw data and human experience. Recognizing emotions' role in decision-making empowers us, turning instinctive reactions into deliberate, insightful actions.

Brain Processes in Emotional Decision-Making

Emotions weave through every decision we make, often quietly guiding us through the tangled web of choices. These emotional threads, spun from intricate brain processes, shape our decisions in ways both subtle and profound. The capacity to make informed decisions isn't merely a product of logic and reason; it's deeply tied to the neural pathways that govern our emotional responses. But what are the brain processes that influence such emotionally charged decision-making?

At the heart of this complex integration of emotion and decision lies the *limbic system*, a cluster of interconnected structures that include the **amygdala, hippocampus**, and parts of the frontal cortex. The amygdala, in particular, functions as the emotional processing center. It plays a crucial role in assessing the emotional significance of stimuli, acting as an emotional filter that determines our responses. When faced with decisions, it evaluates potential threats and rewards, often nudging us toward more intuitive, emotionally-driven choices.

Consider the role of the **prefrontal cortex**, the brain's executive center. It integrates emotional signals with cognitive processes, enabling us to weigh potential risks and benefits. This synthesis between the rational and the emotional isn't a mere balancing act; rather, it's a dynamic interplay that shapes our preferences and influences our judgments. While logic might dictate a certain path, emotional cues from the limbic system can steer us differently, often in ways that align closely with our values and past experiences.

The concept of **emotional tagging** is particularly insightful here. Our brains attach emotional significance—or tags—to the memories and experiences that inform our future choices. These emotional tags help prioritize these memories, guiding decisions by drawing on past emotional outcomes. For example, if a particular choice previously resulted in happiness, the brain tends to favor that option in similar situations moving forward.

Neurotransmitters also come into play, acting as chemical messengers that influence our decision-making processes. *Dopamine*, often celebrated as the "reward" neurotransmitter, affects decision-making by promoting sensation-seeking behaviors and reinforcing actions that bring about rewards. Pleasant outcomes produce a rush of dopamine, encouraging us to repeat the actions that provided those outcomes in the first place. In contrast, neurotransmitters like *serotonin* can temper our impulses, instilling a sense of calm and promoting a balanced emotional approach to decision-making.

Understanding these neural mechanisms opens pathways to harnessing emotional intelligence in decision-making. By recognizing the impact of the amygdala's sensitivity to threat or reward, individuals can develop strategies to mitigate impulsive decisions driven by fear or excitement. Techniques such as mindfulness and self-reflection help slow down the decision-making process, allowing the prefrontal cortex more time to engage in executive reasoning.

The **orbitofrontal cortex** adds another layer of complexity. This area of the brain is heavily involved in evaluating the reward values associated with potential decisions. It considers both immediate outcomes and long-term consequences, effectively balancing short-term emotional gratification with future well-being. This evaluation process becomes essential in scenarios requiring ethical or moral considerations, where emotionally charged outcomes must be weighed against societal norms and personal principles.

Interestingly, the brain's adaptability plays a pivotal role in emotional decision-making. Through a process known as **neuroplasticity**, experiences shape our neural pathways, allowing for adaptive responses to new information and situations. Emotional experiences thus sculpt the brain, influencing future decisions not merely through learned logic but through deeply ingrained emotional biases. This adaptability highlights the potential for intentional change; by cultivating positive experiences and learning from emotional conflicts, individuals can reshape their decision-making landscape.

While much of decision-making is inherently subconscious, consciously tapping into emotional intelligence can enrich the process. Developing a keen awareness of one's emotional triggers and biases allows for more deliberate decision-making. It's about cultivating a present-state consciousness where emotional information is acknowledged without letting it overwhelm the rational aspects of decision-making.

Furthermore, social and cultural contexts shape our emotional responses and thus our decisions. The brain processes these external influences, integrating societal values and emotional norms into personal decision-making. This is especially evident in **group settings**, where emotional decisions are often influenced by social dynamics and the limbic resonance within a group. Navigating such scenarios requires a nuanced understanding of both individual and collective emotional landscapes, ensuring decisions align with both personal values and societal expectations.

The dual-processing model of reasoning—consisting of the fast, intuitive System 1 and the slower, analytical System 2—

also sheds light on the role of emotions in decision-making. System 1 operates largely on emotional intuitions, offering quick, gut-feel responses, while System 2 engages more deliberate thought processes. Balancing these systems is crucial; emotional instincts provide valuable insights but should be evaluated alongside logical reasoning for decisions that balance feeling and fact.

Ultimately, leveraging emotional intelligence in decision-making means embracing our emotional brains as allies rather than adversaries. It's not about suppressing emotions or letting them run unchecked but about using them as guides. In doing so, we align our choices with our authentic selves, tapping into a reservoir of potential that can shape not only our individual lives but also the collective well-being of the communities in which we partake.

Enhancing Decision-Making through Emotional Awareness

Decision-making doesn't occur in a vacuum; it's an intricate dance of logic, intuition, and emotion. Recognizing the discretion of our emotions in this process, especially in professional settings, can redefine the traditional boundaries of cognition and instinct. Weaving emotional awareness into our decision-making toolkit is akin to adding dimensions to a drawing, transforming the flat into the vivid and alive.

Consider the brain as an orchestra where different sections—the prefrontal cortex, the amygdala, and the insula— play their unique instruments. Emotions influence decisions through these neural pathways. The amygdala, often seen as the sentinel of the brain, detects threats and opportunities. In contrast, the prefrontal cortex, our executive center, weighs those cues against a backdrop of logic and future planning. By fostering emotional awareness, we're not only enhancing our ability to identify these signals but also fine-tuning our responses, determining when to heed an instinct and when to question it.

In our professional lives, this can translate into more nuanced, empathetic decisions. For example, a manager faced with a critical decision about team restructuring can tap into their emotional reservoirs. Emotional awareness allows them to consider not just the operational metrics, but the morale and culture of their team. This holistic approach often results in decisions that align better with long-term goals and human values.

Imagine a situation where a business must decide whether to adopt a new technology. Facts and figures might showcase benefits, but the awareness of emotional aversions or enthusiasms within the team can guide the timing and strategy for implementation. It prevents overlooked risks and sidesteps potential resistance by preemptively addressing concerns. Emotional awareness provides a lens to see beyond the immediate tides to the undercurrents shaping them.

But how do we cultivate such emotional awareness in the heat of decision-making? The journey starts with self-reflection, a skill inherently connected to self-awareness and empathy, discussed in previous chapters. Regularly setting aside time to reflect on past decisions—both successful and flawed—helps in discerning the emotional cues that influenced those outcomes. Through these reflections, individuals learn not to suppress these cues but to interpret them wisely, turning emotional noise into decisive signals.

Moreover, practicing mindfulness can be instrumental. Mindfulness doesn't negate emotional input but rather provides a balanced stage where different feelings are acknowledged and understood. Through techniques like deep breathing and focused meditation, professionals slow down their decision-making processes, thereby allowing a more comprehensive view of both emotion and rationality. Such practices enhance emotional clarity and, in turn, facilitate more enlightened choices.

Additionally, engaging with one's emotional intelligence through continuous learning and interaction with diverse professional and personal milieus enriches emotional awareness. This enrichment is like expanding a musical repertoire—each added piece offers new perspectives and deeper understanding. Engaging with different perspectives allows us to anticipate a wider range of emotional reactions, making our decisions more inclusive and considerate.

Efforts aimed at enhancing emotional awareness should also focus on external influences. Social connections and feedback loops, vital aspects of decision-making that involve others, deepen emotional insights. Seeking input from trusted colleagues or mentors can illuminate blind spots one might have, offering alternative viewpoints filtered through different emotional experiences. Such interactions not only validate personal emotions but also bring a collaborative wisdom into the decision-making equation.

On a broader organizational level, nurturing a culture that values emotional input as much as data-driven strategies can forge stronger teams and more innovative solutions. Encouraging open discussions about the emotional impacts of decisions among team members creates an environment where emotional intelligence thrives. This collective emotional awareness acts as a stronghold, bolstering resilience and adaptability in the face of inevitable challenges.

Ultimately, integrating emotional awareness into decision-making is about embracing a more authentic, empathetic approach that resonates at both personal and professional levels. It requires courage to challenge engrained patterns of suppressing emotions in the realm of business logic. Yet, those who do often find themselves not just making better decisions, but also feeling more aligned with the deeper values these decisions reflect.

In summation, enhancing decision-making through emotional awareness enriches the human experience at work, fostering environments where creativity, empathy, and intelligence intersect productively. By understanding and embracing the role of emotions in decision-making, we open doors to a world of unseen connections and possibilities. The decision-making process becomes not just a transactional activity but a transformational journey of growth and harmony with oneself and others.

Chapter 21:
Aging, Brain Health, and Emotional Intelligence

As we age, the landscape of our brains changes significantly, influencing our emotional intelligence along the way. It's a journey where wisdom and emotional understanding can flourish, but only if nurtured with care and intention. Neuroscience shows us that cognitive decline isn't an inevitable part of aging; rather, our brains possess a remarkable capacity for adaptation, thanks to neuroplasticity. This adaptability is crucial in maintaining our emotional intelligence, as the neural pathways governing empathy, self-awareness, and emotional regulation can be reinforced and expanded at any age. With age, emotional intelligence can actually grow deeper, like the roots of an ancient tree, steadying us through life's storms. To harness this potential, regular cognitive exercises, social interaction, and reflective practices become essential. These not only preserve mental acuity but also enhance our ability to navigate the complex emotional landscapes we encounter. Aging, with its symphony of experiences, offers a unique opportunity for growth in emotional intelligence, empowering us to live each moment with greater empathy, understanding, and wisdom.

Emotional Intelligence Changes Over Time

As we traverse the landscape of life, the intricate tapestry of emotional intelligence (EI) unfolds, presenting a fascinating blend of constancy and evolution. Emotional intelligence, a symbiosis of self-awareness, empathy, and emotional regulation, doesn't remain static. It adapts, shifts, and matures, echoing the transformations happening within our brains as we age. Unpacking these changes requires viewing emotional intelligence not just as a skill set but as a dynamic process intimately connected to our life's milestones.

Throughout childhood, emotional intelligence begins its foundation. The brain is a sponge, soaking in experiences, emotions, and social connections. During these formative years, key areas like the prefrontal cortex and amygdala are actively wiring and rewiring, setting the stage for basic emotional competencies. Children are not merely passive recipients of emotions; they're active learners, observationally constructing narratives around feelings they experience and witness.

By the time adolescence rolls around, a significant overhaul is underway in the brain's structure and function. One pivotal change is in the connectivity between the prefrontal cortex and the limbic system, which is responsible for regulating emotions. As this network strengthens, adolescents often find themselves on a roller coaster of emotions. However, this tumultuous period is also ripe for cultivating emotional intelligence. Awareness of interpersonal dynamics deepens, providing fertile ground for empathy and improved self-regulation skills.

Moving into adulthood, our emotional intelligence enters a refinement phase. With the maturing of cognitive abilities, adults can engage more deeply in reflective practices, allowing for nuanced understanding of their emotional landscape. The prefrontal cortex, now fully developed, plays a central role in this process, enabling adults to process complex emotional cues and enhance their decision-making capabilities. A significant body of research suggests that EI peaks at various stages in adulthood, depending on one's lifestyle, challenges, and the degree of emotional introspection they practice.

However, aging brings unique twists to emotional intelligence. In older adults, there is often a noticeable shift towards emotional positivity. Theories suggest this shift may be linked to a life stage focus on maximizing positive experiences and minimizing regrets. Interestingly, the structural and chemical changes within the aging brain, such as the decline in certain neurotransmitters and structural brain changes, can lead to a heightened focus on emotionally rewarding activities and relationships.

Despite the benefits, aging also presents challenges. Cognitive decline, a natural part of the aging process, can affect areas related to EI, making emotional regulation more difficult. However, the plastic nature of the brain offers hope. Engagement in mentally stimulating activities, social connections, and a healthy lifestyle can act as buffers, helping maintain and even enhance emotional intelligence well into older age. Importantly, practices like mindfulness and continued learning can help older adults harness the wisdom of experience, serving as a bedrock for emotional growth and resilience.

The role of experience in this evolution cannot be understated. Life experiences contribute significantly to the development of emotional intelligence. Each challenge faced and navigated not only adds to our emotional vocabulary but enriches the brain's emotional processing circuits. Young adults facing unexpected life scenarios may initially react based on instinctive emotional responses. However, over time and with reflection, they harness these experiences into a more sophisticated EI framework. Similarly, seasoned adults often demonstrate profound empathy and insight, skills honed through navigating myriad life experiences over decades.

Cultural and societal influences also play a role in how EI changes over time. Different cultures emphasize various aspects of emotional intelligence, which can affect how individuals perceive and manage emotions as they age. Societal norms, expectations, and support systems provide scaffolding that can either enhance or hinder the growth of emotional intelligence across the lifespan.

Importantly, gender can affect the trajectory of emotional intelligence over time. Societal norms often shape how emotions are expressed and managed across genders, potentially leading to different pathways in emotional maturation. These nuances provide areas for future research, encouraging a reconsideration of how emotional intelligence is nurtured across different demographics.

Over time, the perception of what constitutes emotional intelligence might expand or shift. In our current era, where emotional intelligence is increasingly valued in personal and professional domains, the conversation around EI is evolving. We are beginning to understand the interplay between social, emotional, and cognitive domains more holistically. This understanding opens the door to a more integrative approach, where emotional intelligence is seen not as separate from cognitive intelligence but as interconnected and complementary.

Finally, there's the notion of continuous improvement. We're never too old to learn new ways to manage and perceive emotions. Lifelong engagement in activities that challenge the brain and emotional capabilities can promote further growth in emotional intelligence. This pursuit of growth forecasts a future where emotional intelligence continues to evolve, reshaping our understanding of how it influences and enriches all stages of life.

The journey of emotional intelligence over time is a poignant reminder of the brain's remarkable adaptability and resilience. Every stage of life offers opportunities for growth, fueled by our experiences, environments, and the choices we make. By illuminating the path of emotional intelligence through the lifespan, we empower ourselves and others to cultivate a fulfilling, emotionally rich life.

Maintaining Emotional Intelligence Through Lifespan

In the journey of life, we pass through various phases, each accompanied by different responsibilities, challenges, and emotional landscapes. But one aspect of our human experience that can anchor us amidst this evolution is emotional intelligence (EI). Like a lifelong companion, EI adapts with us, providing an essential toolkit to navigate both the joys and adversities of life. Maintaining this asset requires deliberate effort across the lifespan.

Our understanding of emotional intelligence suggests that it isn't stagnant; it grows and transforms with age. Much like the brain itself, which continues to change over time, EI has the potential to develop further as we age. This concept aligns with the notion of lifelong learning, where individuals continuously build upon their previous experiences and knowledge. During every decade of life, from early adulthood to the golden years, there are unique opportunities to enhance our emotional intelligence and harness it for more profound personal and professional fulfillment.

The role of emotional intelligence in the aging process is multifaceted. For many, the experiences accumulated over the years enrich their emotional insight. However, maintaining emotional intelligence involves more than just experience; it's about actively engaging with our emotions, reflecting on them, and utilizing them constructively. It requires an ongoing commitment to self-awareness, empathy, emotional regulation, and social skills—core components of emotional intelligence.

In younger years, emotional intelligence can be leveraged to build strong personal and professional relationships. As individuals enter the workforce, the ability to understand and manage emotions becomes crucial for effective communication, teamwork, and leadership. Younger adults often face high-stress environments, whether it's balancing careers, relationships, or personal growth. Here, adapting EI strategies tailored to managing stress, such as mindfulness or emotional regulation techniques, can be exceptionally beneficial. The foundational skills acquired in these formative years can set the stage for lifelong emotional intelligence enhancement.

As we enter midlife, the landscape of emotional intelligence shifts. This stage often involves assessing life priorities and goals. By this age, individuals tend to have a more profound relational depth, understanding the importance of empathy, active listening, and nuanced communication. It's a phase where self-awareness deepens, with many people reflecting on past experiences with a nuanced understanding of their emotional responses. Harnessing this self-reflection can empower individuals to reevaluate their goals and make purposeful changes to better align with their values and desires.

This phase also highlights the importance of flexible thinking. Flexibility in how we respond emotionally to circumstances allows for greater adaptability—an essential trait as people face the realities of aging, such as children growing up, career changes, or the care of aging parents. With greater empathy and understanding, mid-lifers are positioned to provide significant emotional support both professionally and personally, acting as mentors or leaders within their communities.

The later stages of life present unique challenges and opportunities for emotional intelligence development. Aging can bring about physical health concerns and changes in social structures. Yet, it also brings the gift of wisdom and a broad perspective on life's complexities. Older adults often exhibit increased resilience, a quality developed through a lifetime of navigating diverse emotional experiences. This resilience is interwoven with emotional intelligence, providing a robust foundation for tackling the transitions and challenges characteristic of this life stage.

For many seniors, maintaining social connections remains critical in preserving emotional intelligence. Engaging with others—whether through family, community activities, or volunteer work—fosters a sense of belonging and purpose. Through these interactions, seniors continue to practice empathy, understanding, and social engagement, all cornerstones of emotional intelligence. Additionally, there is growing recognition of the role technology can play in aiding social interaction, offering opportunities for engagement even when physical mobility might be restricted.

As humans navigate life's course, emotional intelligence inevitably interlaces with every chapter of their story. By integrating practices that nurture self-awareness, empathy, and emotional regulation throughout the lifespan, individuals can enhance their emotional intelligence. These practices can include engaging in regular self-reflective exercises, maintaining open lines of communication with others, and embracing lifelong learning to adapt to changing emotional landscapes.

The neuroplastic nature of the brain underscores that it's never too late to refine and expand one's emotional capacities. From incorporating mindfulness practices that enhance emotional regulation to fostering environments that encourage social interaction and empathy, numerous strategies are available to support emotional intelligence with age. Ultimately, the ongoing cultivation of emotional intelligence—through conscious, deliberate effort—can lead to a more rewarding, enriched life, one filled with deeper relationships, greater resilience, and a lifelong sense of fulfillment.

Embracing the process of maintaining emotional intelligence across one's lifespan holds promise not only for personal growth but also for contributing positively to the communities we inhabit. Whether it's the support offered to a friend in need, the mentorship provided to a co-worker, or simply the patient listening to a family member, these acts underscore the richness and value of well-maintained emotional intelligence.

Chapter 22:
The Intersection of
Technology and Emotion

In today's fast-paced digital world, technology profoundly shapes our emotional landscape, presenting both challenges and opportunities. While the constant hum of digital chatter can overwhelm our senses and distort emotional responses, it also offers potent tools for enhancing emotional intelligence. Think of the myriad of apps and platforms designed not just to connect us but to offer emotional support, foster mindfulness, and promote mental well-being. Wearables track physiological responses, giving us real-time insights into our emotional states, while artificial intelligence provides personalized feedback on emotional patterns. However, leveraging these technological advancements requires a mindful approach, balancing screen time with real-world interactions to cultivate genuine empathy and connection. In this technological age, the key lies in integrating these innovations with an awareness of mental and emotional health, turning potential digital distractions into tools for empowerment and growth.

Impact of Digital Age on Emotional Development

The digital age is redefining the boundaries of human interaction and emotional growth. This era, marked by the omnipresence of screens and social media, offers both opportunities and challenges to our emotional development. As technology integrates more into our daily lives, it's crucial to understand its profound influence on our emotional health.

In many ways, technology connects us like never before. We can reach out across the globe in seconds, maintaining relationships that might previously have faded. This connectivity can bolster emotional support, allowing people to cultivate friendships and communities without geographical constraints. Online support groups have become a lifeline for many, providing a sense of belonging and an avenue to share personal experiences. For those who previously felt isolated, this newfound accessibility can foster emotional resilience.

However, there's a double-edged sword to this digital connectivity. While it might enhance social bonds, an overreliance on virtual interactions can lead to shallow emotional connections. The loss of face-to-face interactions diminishes the nuanced aspects of communication, such as tone and body language, which are instrumental in developing empathy. The rapid-fire nature of online interaction can also encourage a reduction in patience and depth in conversations, breeding a culture of instant gratification and superficial engagement.

Social media platforms, for all their potential to connect, also have a reputation for instilling anxiety. They often present curated versions of life that can distort reality and amplify feelings of inadequacy and depression. The highlight reels shared by peers can lead individuals to incessant comparisons, fostering negative self-perceptions and insecurities. Emotional intelligence requires self-awareness and confidence, both of which can be undermined by social media's competitive environments.

The 24/7 nature of technology also affects attention spans and the way we process emotions. Information floods in from all directions, making it difficult to focus and reflect deeply on our feelings. Mindfulness, an essential aspect of emotional well-being, can become a rare practice in this high-speed digital climate. Yet, understanding how to maintain such focus amid a barrage of notifications is essential for mental clarity and emotional regulation.

Interestingly, technology isn't all pervasive negativity; it's a powerful tool for emotional development if used mindfully. Educational apps and platforms designed to enhance emotional intelligence are now more accessible than ever. They offer personalized approaches to learning about and managing feelings. With guided meditations, emotional tracking tools, and virtual therapy sessions, technology can provide structured opportunities for emotional growth. These applications demonstrate that while technology can disrupt emotional development, it also has the capacity to foster a deeper understanding of one's feelings.

On a broader scale, virtual reality (VR) technology is breaking new ground in empathy training. VR experiences can simulate scenarios that enable users to "walk in someone else's shoes." This powerful experience can enhance empathy by providing awareness of diverse perspectives, something crucial for emotional mastery. Thus, in leveraging such technologies, people can potentially expand their emotional intelligence in ways traditional methods may never achieve.

Even with its potential, the role technology plays in emotional development largely depends on how it is used. For professionals and lifelong learners seeking to boost emotional intelligence, it is vital to cultivate a balance. By setting boundaries on screen time and prioritizing in-person interactions, one can mitigate the detrimental effects of technology. Mindful consumption, awareness of personal triggers, and taking digital detoxes can optimize mental health amid technological advancements.

Moreover, integrating technology with traditional emotional intelligence practices can offer a balanced approach to personal and professional growth. Journaling apps, goal-setting software, and platforms for feedback can guide individuals in building emotional intelligence systematically. The tools are there for harnessing—but require intentionality in their use.

As technology continues to evolve, so too will its impact on emotional development. We have the opportunity to shape our interactions with technology in ways that enhance, rather than impair, our emotional well-being. The key lies in staying informed of technology's potential effects, utilizing beneficial tools, and maintaining awareness of when digital habits encroach on emotional health. By navigating only with intentionality and balance, we unlock transformative growth, crafting an emotionally intelligent future in an increasingly digital world.

Utilizing Technology for Emotional Enhancement

In the ever-evolving landscape of technology, one might wonder how digital innovations can intersect with the seemingly intangible realm of human emotions. Yet, as technology advances, its potential to influence and enhance our emotional lives is becoming increasingly apparent. At the core of this transformation lies the seamless integration of digital tools into our daily routines, offering professional and personal growth opportunities through increased emotional awareness and intelligence.

Consider the rise of wearable technology. Devices such as smartwatches and fitness trackers are no longer limited to counting steps or monitoring heart rates. They've expanded into the realm of emotional tracking, measuring stress levels, mood fluctuations, and even relaxation states. These wearables employ sensors and algorithms to interpret physiological signals like heart rate variability and skin conductance, allowing users to gain insights into their emotional states. This real-time feedback can be transformative, empowering individuals to recognize stress triggers or moments of calm and adjust their behavior or environment accordingly to maintain emotional balance.

Mobile applications are another massive boon in the journey toward enhanced emotional intelligence. With the advent of smartphones, apps designed to boost emotional wellness are at our fingertips. These digital platforms cover everything from guided mindfulness sessions and stress management techniques to cognitive behavioral therapy exercises and mood journaling. By utilizing data-driven approaches and interactive content, these applications personalize the emotional enhancement process, ensuring that individuals receive strategies tailored to their unique emotional landscapes. Such customization fosters a proactive approach to emotional health, encouraging continuous engagement and learning.

Moreover, artificial intelligence (AI) is playing an increasingly pivotal role in emotional enhancement technologies. AI-driven platforms, equipped with sophisticated natural language processing capabilities, can analyze textual and vocal input to detect sentiment and emotional tones. This analysis aids in understanding and empathizing with one's emotions and those of others. For example, chatbots and virtual assistants can offer empathetic responses, guiding users through challenging emotional terrains or providing support at crucial moments based on detected emotional cues. The possibility of machines contributing to human empathy once seemed far-fetched, yet AI systems are being developed to help individuals navigate complex emotional challenges.

Virtual reality (VR) and augmented reality (AR) present another frontier in emotional enhancement. These immersive technologies offer experiences that can provoke authentic emotional responses, providing a safe space for individuals to practice emotional skills. For instance, VR can simulate social settings or stressful environments, allowing users to practice emotional regulation and communication skills without real-world consequences. Such simulations can be particularly beneficial in therapeutic settings, offering individuals a sandbox to experiment with and refine emotional skills that can then be transferred to everyday interactions.

However, while the potential of technology for emotional enhancement is vast, it's essential to approach this intersection with discernment. Ethical considerations regarding data privacy, the risk of over-reliance on digital tools for emotional well-being, and the need for human connection amidst digital interactions must be carefully navigated. Technology is but a tool—a powerful one indeed—but it should complement human intuition and interpersonal connection rather than replace them.

One must also consider the digital divide's impact on accessing these emotional enhancement technologies. As powerful as these tools can be, they are only available to individuals with adequate digital literacy and access to technology. Efforts must be made to democratize this technology, ensuring everyone can benefit from these advancements, regardless of their socioeconomic status.

In professional settings, emotional enhancement technology can be harnessed to improve workplace dynamics and productivity. By promoting self-awareness and emotional literacy through digital tools, employees can develop better interpersonal relationships and foster a more supportive, empathetic workplace culture. Virtual platforms can facilitate regular emotional well-being check-ins, providing employees with resources to manage stress and maintain motivation. In doing so, organizations not only invest in individual emotional health but also in the collective emotional intelligence of their teams.

Education systems also stand to benefit profoundly from integrating technology with emotion enhancement. By incorporating emotional intelligence training into digital curricula, educators can prepare students for the inevitable challenges of the future, equipping them with the emotional resilience and adaptability necessary in an ever-changing world. Interactive simulations and AI-driven feedback can tailor learning experiences, promoting emotional growth alongside academic achievement.

Technology's impact on emotional enhancement is not restricted to individuals or groups; it also has societal implications. Social networks and digital platforms promote interconnectedness, offering forums for emotional expression and support. They can amplify understanding, empathy, and unity on a broader scale, breaking down geographical and cultural barriers to enable shared emotional experiences.

The interplay of technology and emotion is rich with possibilities for personal and professional growth, but it demands an approach that balances innovation with humanity. As we harness technology to enhance our emotional lives, we must also cultivate digital empathy and emotional literacy to navigate the challenges of an increasingly digital world. When these elements align, technology and emotion can lead to a more connected, understanding, and emotionally intelligent society.

Chapter 23:
Neurotherapy and
Emotional Healing

In the rapidly evolving field of neuroscience, neurotherapy stands out as a groundbreaking approach to emotional healing. It leverages the brain's inherent capacity to change and adapt, offering powerful therapeutic paths for those seeking emotional balance and recovery. Imagine a therapy that taps into the richness of neural pathways, redefining how we view mental health treatment. Advances in brain-based therapies provide practitioners with tools to stimulate neuroplasticity effectively, promoting emotional well-being through precise interventions. Practical applications, such as neurofeedback, help individuals gain insight into their own brain activity, fostering a deeper understanding of the mind-body connection. These innovative techniques empower professionals and lifelong learners to unlock new avenues for personal and professional growth, translating the complex language of the brain into achievable steps for enhancing emotional health.

Advances in Brain-based Therapies

It's an exciting time in the realm of brain-based therapies. As interest in the intersection of neuroscience and emotional healing grows, so does the pool of innovative approaches to treating emotional and psychological challenges. Brain-based therapies leverage cutting-edge neuroscience to tailor interventions that foster emotional healing. These advances mark progress not just for individuals grappling with their emotions, but for professionals aiming to integrate science-backed methods into their therapeutic practice.

One of the most compelling advancements in this field is neurofeedback therapy, an approach that trains individuals to regulate their brain activity. Neurofeedback provides real-time feedback on brainwave patterns, guiding patients to modify their thought processes and reactions. This modality has gained traction in treating conditions like anxiety, depression, and PTSD. Patients often describe the experience as enlightening, revealing previously unknown patterns that contribute to their emotional distress. Neurofeedback aims to give individuals greater control over their emotions and mental states, promoting self-regulation and healing.

Another innovative therapy is transcranial magnetic stimulation (TMS). TMS involves using magnetic fields to stimulate nerve cells in specific brain regions. It's particularly noted for its efficacy in treating major depression, especially in cases resistant to traditional treatments. Unlike medications, TMS targets discrete areas of the brain, reducing the risk of systemic side effects. This ability to localize treatment enhances its appeal as a precise tool in managing emotional disorders. Patients undergoing TMS often report mood improvements and a renewed zest for life, evidencing the potential of brain-centric approaches in reshaping emotional well-being.

Cognitive remediation therapy (CRT) is making waves as another promising advancement. Originally developed for cognitive impairment in schizophrenia, CRT is now being adapted for emotional healing. The therapy focuses on improving cognitive functions like memory, attention, and executive functioning. Enhancing these cognitive abilities allows individuals to process emotions more effectively, impacting overall emotional intelligence. As a result, patients find themselves better equipped to manage emotional upheavals and improve interpersonal relationships.

Crossing over into the realm of the everyday, brain-based self-help tools and apps are democratizing access to effective emotional therapies. Wearable devices that monitor physiological signals such as heart rate and galvanic skin response provide users with data about their stress levels. These insights enable users to adjust their behaviors and thoughts in real-time, fostering emotional regulation. Mobile apps that guide users through practices like mindfulness meditation scripts or cognitive-behavioral techniques are increasingly popular. These tools are rooted in neuroscience principles, offering accessible avenues for emotional mastery.

The therapeutic potential of virtual reality (VR) in treating emotional disturbances is another advance worth noting. VR creates immersive environments that serve as controlled settings for exposure therapy and anxiety treatments. Patients can interact with simulations that mimic anxiety-inducing scenarios in a safe way, helping them develop coping strategies and reduce anxiety responses. VR's unique ability to safely transport individuals to highly specific environments provides therapists with additional control over the therapeutic process, boosting effectiveness.

Equally transformative is the role of personalized medicine in brain-based therapies. The use of genetic, molecular, and neuroimaging information allows for more personalized treatment approaches. Understanding the unique neurological, genetic, and biological makeup of patients empowers therapists to craft personalized interventions. This tailored approach enhances the effectiveness of treatments by aligning therapeutic strategies with individual needs.

Moreover, recent advances in brain imaging technology have provided unprecedented insights into the neural circuits involved in emotional processing. Functional MRI (fMRI) and PET scans enable scientists to observe brain activity in real-time, forging direct links between neurological changes and emotional states. These insights offer therapeutic pathways, helping professionals to design interventions that specifically target malfunctioning brain areas.

Advancements in understanding the microbiome-gut-brain axis also offer new dimensions in neurotherapy. Research increasingly shows that gut health significantly influences mental well-being. Probiotics, diet modifications, and lifestyle adjustments are modern tools used to enhance emotional health. This approach underscores the interconnectedness of physical and emotional health, prompting a holistic perspective in therapy design.

The therapeutic landscape is further enriched by developments in the field of psychedelics, explored under controlled, therapeutic settings. Early research suggests substances like psilocybin and MDMA hold promise in treating PTSD, anxiety, and depression. These substances, when administered and monitored in a therapeutic context, may facilitate deep emotional breakthroughs by altering functional connectivity in the brain.

Overall, advancements in brain-based therapies signify a hopeful turn in emotional healing. For practitioners and individuals alike, these innovations provide new strategies to support emotional health. As neuroscience continues to peel back layers of the brain's mysteries, the prospect for even more targeted, effective therapeutic interventions grows. Here lies empowerment—equipping individuals with scientifically grounded tools to navigate and transform their emotional landscapes with confidence and hope.

Practical Applications of Neurotherapy for Emotional Health

Neurotherapy, with its focus on manipulating and enhancing brain function, is gaining traction as a forward-thinking approach to emotional health. By using techniques that directly influence brain activity, practitioners aim to reshape the way individuals experience and manage their emotions. This method promises significant benefits, especially for those grappling with emotional challenges in today's fast-paced world.

It's essential to understand that neurotherapy isn't a single technique but a collection of interventions aimed at optimizing brain function. Techniques like neurofeedback, transcranial magnetic stimulation (TMS), and cranial electrotherapy stimulation (CES) have emerged as key players in this domain. Neurofeedback, for instance, involves training the brain to self-regulate by providing real-time feedback on neural activity. It's akin to teaching the brain how to respond more effectively to emotional stimuli.

Although this might sound like something out of science fiction, neurofeedback has been around for decades. In mental health settings, this technique has been used to address a variety of conditions including anxiety, depression, and PTSD. By helping the brain learn new patterns of activity, neurofeedback can foster improved emotional resilience. The changes aren't just theoretical. People report feeling calmer and more centered after sessions, suggesting that the adjustments in neural activity translate into tangible emotional benefits.

Transcranial magnetic stimulation is another exciting development in the application of neurotherapy. Unlike neurofeedback, which relies on the patient's active participation, TMS directly stimulates the brain using magnetic fields. This method has shown promise in treating depression, particularly in cases where traditional therapies haven't been effective. By targeting specific brain regions involved in mood regulation, TMS can alter neural pathways and enhance emotional well-being.

One of the more user-friendly methods of neurotherapy is cranial electrotherapy stimulation. It's like an electronic massage for the brain. CES devices deliver small electric currents to the cranium, which may help stabilize emotional states by impacting neurochemical activities. Studies have shown that CES can reduce symptoms of anxiety and depression, possibly by encouraging the production of mood-regulating neurotransmitters like serotonin and dopamine.

What makes neurotherapy particularly appealing is its potential for customization. Because emotional responses are deeply personal and varied, interventions can be tailored to meet individual needs. No two brains are identical, and neurotherapy recognizes this by offering personalized treatment plans. Imagine receiving therapy that precisely targets your unique neural patterns, fine-tuning them for optimal emotional performance. It's a transformational approach that holds great promise for individual well-being.

Furthermore, the applications go beyond treating disorders. Neurotherapy can be utilized to enhance emotional intelligence, boost focus, and improve stress management. In high-pressure environments like corporate settings, where emotional acuity and resilience are crucial, neurotherapy can serve as a powerful tool for personal development. High-performers often look for anything that gives them an edge, and neurotherapy could be just that.

Emotional regulation is not just about controlling negative emotions; it's also about cultivating positive ones. Neurotherapy can aid in reinforcing positive emotional experiences and reinforcing neural pathways associated with joy and contentment. Over time, the brain becomes more adept at accessing these states, contributing to a more consistently positive outlook. The implications for long-term well-being and quality of life are profound.

From a scientific perspective, the theoretical underpinnings of neurotherapy are supported by our growing understanding of neuroplasticity. The brain's ability to remodel itself in response to experiences and interventions is at the core of why neurotherapy can be effective. By encouraging specific patterns of brain activity, these techniques leverage the brain's natural capacity for change, promoting healthier emotional responses.

In practice, adopting neurotherapy involves a commitment to regular sessions, as well as a willingness to engage in parallel lifestyle changes. It's not a stand-alone cure-all but rather part of a holistic approach to mental health. Incorporating practices like meditation, exercise, and healthy nutrition can complement and enhance the benefits achieved through neurotherapy, creating a comprehensive strategy for emotional health.

Critically, as with any emerging field, it's important to approach neurotherapy with a discerning mind. While the science is promising, it should be conducted by qualified professionals with a thorough understanding of brain function. Rigorous evaluation and ongoing research are needed to keep refining these techniques and understand long-term outcomes.

In summary, the practical applications of neurotherapy for emotional health are as diverse as they are promising. From aiding individuals with mental health disorders to enhancing emotional intelligence across various contexts, neurotherapy offers a cutting-edge avenue for those seeking to improve their emotional lives. With continued research and application, the potential to reshape emotional health on a grand scale is not just a possibility; it may very well be the future of mental health care.

Chapter 24:
Emotional Intelligence
in Leadership

In the dynamic world of leadership, emotional intelligence serves as a cornerstone for truly effective influence and guidance. Leaders who master the art of emotional intelligence can navigate the complexities of interpersonal relationships and strategic decision-making with grace and empathy. The neural intricacies that underpin this form of intelligence allow leaders to not only recognize their own emotions but also to perceive and influence the emotional states of their teams. By embracing techniques rooted in neuroscience, leaders can cultivate essential skills such as empathy, self-regulation, and motivation, thus fostering an environment of trust and collaboration. Emotionally intelligent leaders are adept at balancing analytical thinking with emotional insight, creating a harmonious workplace where creativity and innovation thrive. Indeed, emotional intelligence is not merely an asset in leadership—it's an imperative, enabling leaders to inspire, support, and sustain resilience in themselves and others. This chapter will illuminate how these attributes not only enhance personal growth but also propel organizations toward meaningful success.

Brain Traits of Effective Leaders

In the complex world of leadership, certain brain traits distinguish successful leaders who harness emotional intelligence to its fullest potential. The intricate dance between cognitive capabilities and emotional understanding forms the backbone of what we term 'effective leadership.' This delicate balance is grounded in neuroscience and revolves around our brain's ability to process and adapt across diverse situations, making it a focal point for anyone seeking to improve their leadership skills.

An effective leader is not just someone with charisma or technical expertise; instead, they possess a refined ability to understand and manage emotions both within themselves and in others. Central to this capability is the prefrontal cortex, the brain region responsible for executive function. It manages decision-making, complex cognitive behavior, and moderating social conduct. Leaders with well-developed executive functions demonstrate the ability to plan strategically, adapt to changing circumstances, and chart a course that aligns with long-term objectives.

Empathy, another critical trait of effective leaders, originates from specific neural circuits in the brain, primarily the mirror neuron system. These neurons activate our understanding of emotions and intentions behind others' actions, making them crucial for empathetic communication. When leaders actively engage these neural pathways, they cultivate a deeper connection with their team, enhancing trust and cooperation.

Complementing empathy is the ability to maintain social bonds and emotional regulation, which is significantly shaped by the amygdala and the hippocampus. The amygdala plays a vital role in processing emotions and threats, while the hippocampus is involved in forming new memories and connecting emotions to those experiences. Leaders who manage to keep their amygdala in check and use their hippocampus for reflective thinking can respond to emotional triggers with calm and considered actions.

Another pivotal trait is resilience, supported by the brain's ability to adapt through neuroplasticity. Resilience allows leaders to bounce back from setbacks and to view challenges as opportunities for growth. Neuroplasticity, the brain's remarkable capacity to reorganize itself by forming new neural connections, empowers leaders to continually refine their emotional responses and problem-solving strategies. This adaptability is critical for leaders, as it allows them to thrive amidst evolving circumstances.

The trait of self-awareness is crowned among brain traits that support effective leadership. Neuroscientists link self-awareness to regions such as the anterior cingulate cortex and the insular cortex, both of which facilitate introspection and self-monitoring. Leaders with heightened self-awareness are better equipped to recognize their emotional states, understand the impact of their behavior on others, and control their reactions to maintain harmony in their teams.

Decision-making, inherently influenced by emotions, is another cornerstone of effective leadership, often seen as a reflection of a leader's rational brain interacting with their emotional circuits. The interplay between the rational prefrontal cortex and emotional regions like the amygdala ensures that decision-making is both informed and empathetic. Leaders who excel in this area are capable of making decisions that are not just logical but also considerate of the emotional dynamics in their environment.

Curiosity and the ability to foster creativity within a team are also linked to effective leadership. These traits engage the brain's reward system, particularly the ventral striatum and the nucleus accumbens, promoting exploratory behavior and innovative thinking. When leaders encourage curiosity and creativity, they ignite a passion for learning and problem-solving in their teams, leading to dynamic and forward-thinking organizational cultures.

In aligning these traits, emotional intelligence forms the crux of brain functions that drive effective leadership. Emotional intelligence is about being smart with feelings—recognizing, understanding, and managing our emotions and the emotions of others. Leaders who exhibit high emotional intelligence foster environments where team members feel valued, understood, and empowered to contribute their best work.

Motivated by neuroscience, leaders can consciously nurture these brain traits, aligning personal development with scientific understanding. Embracing ongoing learning and self-reflection becomes a gateway to cultivating these abilities. Tools such as mindfulness, feedback loops, and adaptive thinking strategies play essential roles in enhancing these neural capacities, resulting in more effective, emotionally intelligent leaders.

The journey to leadership mastery is not one of destination but rather a continuous progression, informed by the brain's inherent ability to learn and adapt. Effective leaders serve as architects of their own cognitive development, intentionally sculpting their neural landscapes to meet the challenges of modern leadership with grace and empathy. Through understanding and leveraging these key brain traits, one can foster a leadership style that inspires, motivates, and transforms both individuals and organizations.

Cultivating Leadership Skills through Emotional Intelligence

In the modern landscape of organizational dynamics, emotional intelligence has emerged as a cornerstone of effective leadership. Leadership isn't merely about delegating tasks or making strategic decisions. At its core, it requires the capacity to recognize, understand, and manage one's emotions while also navigating the emotions of others. An emotionally intelligent leader exudes an understanding that transcends traditional intellectual capabilities.

Research highlights that emotional intelligence accounts for nearly 90% of what sets high-performing leaders apart from their peers. It's not merely a soft skill; it's a strategic asset. Leaders with high emotional intelligence are adept at managing their emotions, which in turn helps them lead their teams more effectively. But how can one cultivate these essential skills? Through a deeper understanding of the components of emotional intelligence, one can embark on this transformative journey.

The first step in cultivating emotional intelligence in leadership is developing self-awareness. Leaders should start by turning the lens inward, engaging in introspective practices that allow for the identification of emotional triggers and understanding how these emotions influence decision-making processes. By recognizing the nuances of their emotional responses, leaders can gain greater control and respond more effectively in diverse situations.

Self-regulation is the next vital competency. This involves the ability to manage one's emotional responses and maintain composure even under pressure. Leaders need to practice self-discipline, remaining resilient in the face of adversity, and avoiding impulsive decisions that could lead to potentially adverse outcomes. Reflective practice and mindfulness can be powerful tools in this regard, helping leaders stay grounded and focused.

An essential aspect of emotional intelligence is motivation—a drive that goes beyond mere external rewards. Leaders with high emotional intelligence possess an intrinsic passion for achieving goals, leading change, and inspiring others. They display a commitment and a visionary mindset that propels both individuals and teams toward collective aspirations. By aligning personal values with organizational goals, leaders can foster an environment of authenticity and drive.

Empathy, perhaps one of the most discussed aspects of emotional intelligence, is about truly understanding and sharing the emotions of others. For leaders, this means listening actively, being attuned to verbal and non-verbal cues, and appropriately responding to team members' needs and concerns. Empathetic leaders build trust and a sense of belonging among their teams, leading to enhanced collaboration and increased morale. Empathy is not innate for everyone; effective training and awareness exercises can sharpen this skill.

Social skills, the final pillar, are imperative for orchestrating change and managing conflict within organizations. Highly emotionally intelligent leaders excel in communication and relationship management. They understand the nuances of social networks and leverage these relationships for collective success. Effective communicators are often confident negotiators, capable of resolving disputes and guiding teams through challenging times.

Moreover, emotional intelligence in leadership extends to creating emotionally healthy workplace cultures. Leaders are tasked with modeling emotional intelligence, inspiring such traits in others, and fostering an environment where open communication and mutual respect are the norm. This cultural shift enhances team cohesion and supports the mental well-being of employees, ultimately leading to improved organizational performance.

Cultivating emotional intelligence is a continuous journey. It requires dedication, self-reflection, and an openness to learning. Leaders can benefit from feedback mechanisms, coaching, and training programs that focus on emotional intelligence competencies. These resources can provide valuable insights into personal and professional growth, enabling leaders to harness neuroscience for their development.

It's also crucial for leaders to foster emotional intelligence by understanding the specific psychological and neurological processes involved. By comprehending the brain's role in emotion and decision-making, leaders can enhance their capacity to respond to a wide array of situations. Neuroscience-backed techniques, such as mindfulness and cognitive reappraisal, can significantly augment a leader's emotional regulatory capabilities.

For those ready to embark on this path, the key is to begin by setting tangible goals for emotional intelligence development. These goals could involve seeking feedback from peers, examining emotional responses in critical situations, or practicing emotional regulation techniques. Over time, these practices can lead to a deepened understanding of emotions and their pivotal role in leadership efficacy.

Ultimately, emotionally intelligent leadership doesn't just elevate the individual leader; it transforms organizations. By wielding the power of emotion with skill and foresight, leaders can inspire a sense of purpose, foster innovation, and drive sustainable change. As leaders grow in emotional intelligence, so too do the institutions they guide. They become architects of environments in which employees thrive and potential is unleashed.

Emphasis on emotional intelligence heralds a new era of leadership—one defined by connection, compassion, and profound understanding, opening a window to unprecedented opportunities for organizational and personal transformation.

Chapter 25:
Future Directions in
Emotional Neuroscience

As we explore the exciting frontiers of emotional neuroscience, new trends are poised to revolutionize our understanding and application of these insights for personal and professional growth. Advances in technologies like artificial intelligence and machine learning are not only broadening the horizons of research but also democratizing access to emotionally intelligent interventions. Emerging evidence highlights the potential of personalized brain-computer interfaces to optimize emotional well-being, offering highly tailored strategies for managing mental states that reflect our unique neurological landscapes. The blending of neurofeedback techniques with everyday technologies is creating real-time opportunities for emotional calibration, encouraging individuals to refine their emotional landscapes with precision. Simultaneously, advances in genetic neuroscience promise to unravel the intricate interplay between our emotional capacities and genetic codes, opening new pathways for personalized emotional intelligence. The intersection of cross-disciplinary research continues to inspire innovative approaches for harnessing the brain's incredible plasticity, empowering us to navigate life's emotional challenges with resilience and astuteness. As we stand on the brink of these advancements, the imperative is clear: embracing the integration of cutting-edge neuroscience into our daily lives can truly transform our emotional experiences and elevate our interpersonal engagements.

Emerging Trends in Neuroscience and Emotion

In recent years, the intersection of neuroscience and emotion has become an exciting frontier for researchers and practitioners alike. The dynamic field is ushered forward by innovations that not only reshape our understanding of how emotions are processed in the brain but also how these insights can be applied to enhance emotional intelligence and interpersonal skills. As we stand at the cusp of this fascinating evolution, it's crucial to explore some emerging trends that are likely to drive the future of emotional neuroscience.

One remarkable trend is the use of advanced neuroimaging techniques. Functional Magnetic Resonance Imaging (fMRI) and Positron Emission Tomography (PET) have paved the way for visualizing the brain in real-time during emotional experiences. Now, newer modalities, like optogenetics and chemogenetics, are offering unprecedented precision in mapping emotional processing in the brain. These techniques allow researchers to activate or deactivate specific neural circuits and assess the subsequent impact on emotion and behavior, providing transformative insights into the biological underpinnings of emotion.

Concurrently, the role of artificial intelligence (AI) and machine learning in emotional neuroscience is gaining momentum. Through massive data analysis, AI can identify patterns in how different brain regions interact during emotional states. This is creating possibilities for personalized interventions aimed at modifying emotional responses. Imagine AI algorithms that could analyze brain scans to predict emotional dysregulation in individuals and suggest tailored neuroscience-backed strategies to counteract these challenges. Such tools are poised to become invaluable in both clinical and non-clinical settings.

Another emerging area is the study of gut-brain interactions and their impact on emotions. The gut, often referred to as the "second brain," communicates with the central nervous system through the intricate vagus nerve pathway. Recent research suggests that gut health, influenced by diet and microbiota composition, can significantly affect mood and emotional resilience. This burgeoning field is not only enriching our comprehension of emotional health but also emphasizing the importance of holistic intervention strategies that include dietary and lifestyle changes.

Additionally, the field of epigenetics is introducing new perspectives on how environmental and psychological factors can modify gene expression related to emotion. Understanding the epigenome's role in shaping emotional responses can illuminate why individuals react differently to similar emotional stimuli. This insight is crucial as it opens pathways for interventions that don't just alter behavior temporarily but initiate changes at the genetic level to promote lasting emotional well-being.

Emerging technologies in wearable devices are also playing a significant role in this landscape. Devices that can monitor physiological signals such as heart rate variability and galvanic skin response provide real-time data on emotional states. With the integration of these sensors with smartphone applications, individuals can receive immediate feedback and recommendations to manage their emotional responses effectively. This type of biofeedback is a game-changer for enhancing emotional intelligence, allowing for a proactive rather than reactive approach to emotional regulation.

Furthermore, there's a growing trend in the neuroscience community towards understanding the cultural and social dimensions of emotions. While the biological basis of emotion is universal, the expression and regulation of emotion can vary significantly across cultures, affecting interpersonal dynamics and emotional intelligence. By exploring these cultural aspects, scientists can develop culturally sensitive models of emotional processing, which in turn inform better interventions and communication strategies across a diverse global population.

The exploration of neuroplasticity continues to be a core trend, emphasizing that the brain's structure and function can be continually reshaped throughout life. This concept is particularly empowering for emotional development, suggesting that interventions like cognitive-behavioral therapy, mindfulness practices, and even certain physical exercises can restructure emotional circuitry. Such interventions can foster emotional resilience and regulation, vital components of emotional intelligence.

In line with neuroplasticity, the concept of mental fitness is gaining traction. Just as physical fitness requires consistent practice and training, mental fitness involves systematic approaches to strengthen emotional muscles. Techniques like gratitude exercises, visualization, and emotional journaling are appearing in research as effective means to enhance positive emotional states and fortify mental resilience. The idea is to not only cope with adversity but thrive through mental agility and emotional strength.

Last but not least, the ethics of emotional neuroscience are becoming increasingly pertinent. As technology allows deeper insights into our emotional lives, it brings questions about privacy, consent, and the potential misuse of neuroscientific insights. Stakeholders are beginning to draft guidelines and policies to ensure that advances in emotional neuroscience benefit society equitably and ethically. Being aware of these considerations is essential for anyone involved in the application of neuroscience to emotional development.

The horizon of emotional neuroscience is brimming with potential. These trends indicate a seismic shift towards an integrative approach where biological, technological, and socio-cultural dimensions converge to revolutionize our understanding of emotions. As we continue to embrace these innovations, we hold the key to unlock new levels of emotional awareness, intelligence, and mastery, paving the way for enhanced interpersonal relationships and personal growth. As professionals and lifelong learners, staying abreast of these developments not only empowers us but sets the stage for the transformative application of emotional neuroscience in our lives and communities.

Integrating Future Insights into Personal Growth

As the field of emotional neuroscience continues to evolve, it's evident that understanding how our brains influence emotions can significantly enrich our personal and professional lives. Emerging insights from neuroscience offer a wealth of potential strategies to boost emotional intelligence and foster more profound personal growth. By integrating these insights into our daily practices, we're not just spectators of the future of brain science but active participants shaping our emotional landscapes.

One exciting area of future research lies in neurofeedback and its application to emotional self-regulation. Neurofeedback technology provides real-time data about brain activity, empowering individuals to make conscious adjustments that can lead to improved emotional responses. Imagine having the ability to train your brain to respond more calmly to stress or to enhance your empathy during interactions. By incorporating these cutting-edge tools, individuals become more attuned to their neurological states, managing them with a level of precision that was once unthinkable.

Another promising direction involves personalized emotional intelligence training programs. These programs can be tailored to an individual's unique neural patterns, optimizing their emotional growth in ways that are specific to their needs. As our understanding of the brain's plasticity grows, customized approaches will likely become the norm, offering unprecedented opportunities for individuals to enhance their emotional skills efficiently. This bespoke customization means that rather than a one-size-fits-all approach, each person can embark on a journey finely tuned to their neural signature.

Moreover, emerging technologies in virtual reality (VR) and augmented reality (AR) are poised to transform the landscape of emotional training. These technologies can simulate real-world emotional challenges in safe, controlled environments, allowing individuals to practice emotional responses without real-world consequences. Engaging with these immersive experiences enhances empathy, reduces anxiety, and builds emotional resilience, preparing individuals for real-life scenarios with heightened emotional prowess.

Artificial intelligence (AI) is another frontier intersecting with emotional neuroscience. AI-driven platforms are being developed to help in fostering emotional awareness by analyzing and providing feedback on emotional patterns and interactions. These tools can guide individuals toward better understanding their emotions and the emotions of others, leading to improved interpersonal relationships and personal growth. By leveraging AI, users gain insights that might otherwise remain elusive in traditional emotional learning contexts.

Furthermore, advancements in the understanding of the gut-brain axis offer intriguing implications for emotional health. This research underscores the impact of nutrition and gut health on emotional states, suggesting new dietary approaches that could enhance emotional well-being. A future where diet modifications are tailored to bolster emotional states could become a reality, making nutritional neuroscience a critical component of personal growth strategies.

The role of genetics in emotions is another burgeoning area of study. Future directions in genetic research may allow us to pinpoint genetic predispositions to certain emotional tendencies. By understanding these predispositions, individuals could work proactively to mitigate negative emotional patterns and strengthen positive ones. Genetic insights would thereby augment the arsenal of tools available for personal growth, enabling more targeted actions to cultivate emotional resilience and intelligence.

Social robotics, which involves the use of robots designed to interact with humans on a social level, also holds potential. These robots could become companions or facilitators in emotional development, providing real-time feedback and helping individuals practice emotional skills in a non-judgmental setting. The notion of learning and refining emotional intelligence alongside an "emotion coach" that's available anytime might become a cornerstone of personal development.

Engagement with cross-disciplinary studies is also expanding, providing fertile ground for novel approaches to personal growth through emotional neuroscience. Collaborations between neuroscientists, psychologists, educators, and technologists can yield innovative frameworks for emotional learning. Such collaboration enhances broad understanding, integrating diverse insights into comprehensive strategies for emotional development.

Crucially, the ethical considerations surrounding these advancements cannot be overlooked. As we delve deeper into the potentials offered by neurotechnology and genetic insights, it's vital to maintain a dialogue on the ethical implications. Respecting individual privacy, ensuring informed consent, and safeguarding against potential misuse of sensitive data are paramount. Upholding ethical standards ensures that these powerful tools are used to support and enhance human well-being.

Ultimately, by integrating future insights from emotional neuroscience, we're embarking on a transformative journey.

This journey is characterized by enhanced self-awareness, greater emotional resilience, and deeper social connections. As professionals and lifelong learners, adopting these insights equips us with the ability to navigate the complexities of modern life with more purpose and fulfillment.

As we continue to embrace these cutting-edge discoveries, the future promises not only a deeper understanding of our emotional lives but also a wealth of strategies to bolster our well-being. By weaving these insights into the fabric of our everyday experiences, we're not just reacting to change but actively cultivating our emotional landscapes, achieving personal growth that is as dynamic and varied as the neurobiological foundations on which it rests.

Conclusion

As we reach the close of our journey through the intricate landscapes of emotional neuroscience, it's fitting to reflect on the transformation that the merge of science and practice affords us. We've traversed the depths of neural pathways, discovering how brain structures intertwine with our emotions, and we've seen the potential for these insights to catalyze personal and professional evolution. It's more than a mere accumulation of knowledge; it represents an awakening to the power within us to reshape our emotional and cognitive landscapes.

Embracing the principles of neuroplasticity, we uncover a terrain ripe for rewiring and growth. Our brains, far from being static, have the capacity for renovation at each stage of life. This adaptability means that we can cultivate not only resilience but new patterns of emotional intelligence that serve us better. Empowering ourselves with this understanding shifts us from passive receivers of emotion to active architects of our emotional realities. Imagine leveraging your brain's inherent adaptability to sculpt an emotionally intelligent and resilient existence.

Throughout the chapters, one theme becomes abundantly clear: the journey to emotional mastery demands intentionality. Whether through dedicated mindfulness practices, engaging in activities that promote neurochemical balance, or crafting environments that foster empathy and connection, the science is unequivocal—our actions matter. The decision to harness these strategies aligns our neural functions with our aspirations, leading to profound shifts in how we experience and navigate the world.

It's also important to recognize the role of external influences, from nutrition and sleep to social connections and digital technology, in molding our emotional well-being. These elements, when balanced adeptly, provide a framework that supports our emotional goals. However, understanding these relationships underscores another crucial concept: agency. Equipped with knowledge, we possess the power to make choices that align with our desired emotional states. Every meal, every night of rest, and every relationship is a thread in the fabric of our emotional tapestry.

Moreover, as we consider the future of emotional neuroscience, we're reminded of the perpetual evolution of the field. The rapid advancements in technology and research promise even deeper insights into the emotional brain. Such developments not only enhance but will redefine how we understand and interact with our emotions. And while the future holds uncertainty, it undeniably holds potential. By staying informed and open to integrating emerging insights, we prepare ourselves for continuous growth.

In leadership, emotional intelligence undoubtedly emerges as crucial. Brain-based strategies for leading with empathy, vision, and efficacy create ripple effects—impacting teams, organizations, and even industries. Leaders armed with emotional intelligence not only steer success but foster environments where others can develop their own emotional capacities. The interrelationship between leadership and emotional intelligence invites us to envision a future of more emotionally aware workplaces.

To every professional and lifelong learner seeking growth: the synthesis of neuroscience and emotion offers a toolkit for mastering interpersonal skills and enhancing emotional intelligence. These scientific insights are not static facts but living practices, meant to be woven into the fabric of daily life. Your commitment to embracing these tools reflects a dedication not merely to personal enhancement but to enriching the collective human experience.

The voyage depicted within these pages is not an end but a beginning. It's an invitation—a call to harness the extraordinary capabilities nestled within your brain for continued emotional exploration and empowerment. To quote a timeless insight, the only constant is change. But with change comes opportunity, and the interplay between neuroscientific advances and emotional wisdom sets the foundation for a world where we are ever-evolving beings moving through life's complexities with greater ease and grace.

In closing, consider this work a map and a companion for your journey. It's a guide you can revisit, draw from, and expand upon as you continue to evolve in emotional understanding. The increase in emotional intelligence isn't a final destination but an ongoing process. Embarking on this journey signals a commitment to becoming a more aware, empathetic, and emotionally intelligent version of yourself, equipped to inspire and uplift those around you.

Appendix A:
Additional Resources
for Emotional Mastery

In our journey of understanding the intricate dance between emotions and the brain, it's essential to have a repository of resources that bolster continued learning and application. This appendix serves as a guidepost, offering a curated collection of tools, books, and digital platforms to help further enhance your emotional intelligence and mastery. Here, we connect the dots from the scientific insights laid out in the previous chapters to practical resources you can explore and integrate into your personal and professional life.

1. Books and Literature

"Emotional Intelligence 2.0" by Travis Bradberry and Jean Greaves: A practical book that provides real-world strategies for increasing your EQ, which pairs well with our discussions in emotional intelligence.

"Brain Rules" by John Medina: This book breaks down 12 principles for surviving and thriving at work, home, and school, spotlighting how our brains really work.

"The Emotional Life of Your Brain" by Richard Davidson: Focused on the emotional styles impacted by neural patterns, this book offers compelling insights that parallel our exploration into brain-emotion connections.

2. Online Courses and Workshops

Coursera - "Foundations of Positive Psychology": This course from the University of Pennsylvania dives into the concepts of positive emotions, which aligns with our chapter on positive thinking and emotion.

edX - "The Science of Happiness": Explore the roots of a happy and meaningful life, uncovering keys to emotional health and resilience.

Mindful Schools - "Mindfulness Fundamentals": An excellent resource that supports our mindfulness discussions by providing practical steps to develop a mindfulness practice.

3. Digital Tools and Apps

Headspace: A well-known app that offers guided meditation exercises to cultivate mindfulness, thereby assisting in emotional regulation.

Calm: This app can serve as a hands-on tool to manage stress and enhance emotional balance, echoing techniques we've explored.

Moodpath: As a mental health companion, Moodpath checks your mood daily, prompts reflection, and provides access to psychological exercises that align with emotional insight themes.

4. Popular Podcasts and Videos

The Happiness Lab with Dr. Laurie Santos: This podcast presents stories and science to change the way we think about happiness.

"TED Talks" on Emotional Intelligence: Engage with a series of talks that put a spotlight on emotional management and interpersonal skills.

Huberman Lab - "Science Focused Tools for Everyday Life": Offers in-depth insights into neuroscience that are application-focused.

5. Workshops and Training Programs

Emotional Intelligence Training: eLearning and Instructor-Led Training Programs: For educators and professionals seeking immersive training experiences.

Mindful Leader's "Mindful Leadership Summit": Attend workshops that introduce mindfulness within leadership frameworks.

The resources listed above are designed to complement the strategies and theories explored within this book, offering numerous avenues for expanding your prowess in emotional intelligence. The arena of emotional mastery is vast, and these tools aim to contribute to your ongoing evolution, propelling you towards enhanced personal and professional fulfillment. With these assets at your disposal, we hope you're encouraged to further pursue this path of cognitive and emotional discovery.